Introduction of Industrial Minerals

Introduction of Industrial Minerals

George Beckett

Oxford Book Company
Jaipur ▪ New Delhi

Published By

Oxford Book Company

Regd. Off. "Bony Residency", Gate No. 2
Opp. Tilak Public School, Vishweswariya Nagar,
Gopalpura Road, Jaipur - 302018, India
Ph: 0141 - 2761280, Telefax: 2761381
email - oxfordbookjpr@gmail.com
website - oxfordbookcompany.com

Delhi Office: 102, 1st Floor, Satyam House,
4327/3, Ansari Road, Darya Ganj,
New Delhi - 110002, India
Ph: 011-45652440

Distributed by
KEYA INTERNATIONAL PRESS LLP
E-mail: keyainternationalpressllp@gmail.com

First published in India, 2024

ISBN : 9789355241023

Typeset by Shivangi Computers, Jaipur

Printed and bound in India by Replika Press Pvt. Ltd.

Contents

Preface

Industrial minerals are geological materials which are mined for their commercial value, which are not fuel (fuel minerals or mineral fuels) and are not sources of metals (metallic minerals). They are used in their natural state or after beneficiation either as raw materials or as additives in a wide range of applications. Typical examples of industrial rocks and minerals are limestone, clays, sand, gravel, diatomite, kaolin, bentonite, silica, barite, gypsum, and talc. Some examples of applications for industrial minerals are construction, ceramics, paints, electronics, filtration, plastics, glass, detergents and paper. In some cases, even organic materials (peat) and industrial products or by-products (cement, slag, silica fume) are categorized under industrial minerals, as well as metallic compounds mainly utilized in nonmetallic form (as an example most of the titanium is utilized as an oxide TiO_2 rather than Ti metal). In some cases, organic substances (such as peat) are categorized as industrial minerals. This case, however, violates the technical definition of a "mineral." A substance must be a mineral before it may be categorized as an industrial mineral. Peat is a geologic substance mined for its economic value but that does not mandate that it be referred to as a mineral or as an industrial mineral.

The evaluation of raw materials to determine their suitability for use as industrial minerals requires technical testwork, mineral processing trials and end-product evaluation; free to download evaluation manuals are available for the following industrial minerals: limestone, flake graphite, diatomite, kaolin, bentonite and construction materials. These are available from the British Geological Surveyexternal link 'Industrial Minerals in BGS' with regular industry news and reports published in Industrial Minerals magazine. The mineral industry is the branch of industry responsible for the exploitation of minerals from soil deposits. This is achieved by mining (through underground excavations or open workings), but also by processing plants. Products of mineral industry include various building materials, such as rocks (ex. granite), but also cement, glass and ceramics. The mineral industry of Africa is one of the largest mineral industries in the world. Africa is the second biggest continent, with 30 million km^2 of land, which implies large quantities of resources. For many African countries, mineral exploration and production constitute significant parts of their economies and remain keys to future economic growth. Africa is richly endowed with mineral reserves and ranks first or second in quantity of world reserves of bauxite, cobalt, industrial diamond, phosphate rock, platinum-group metals (PGM), vermiculite, and zirconium. Gold mining is Africa's main mining resource.

The Central African Mining and Exploration Company (CAMEC), one of Africa's primary mining enterprises, is criticized for its unregulated environmental impact and minimal social stewardship. In the Spring of 2009, retired British cricket player Phil Edmonds' assets were seized by the United Kingdom's government due to CAMEC's illicit association with self-appointed Zimbabwean President Robert Mugabe. CAMEC recently sold 95.4% of its shares to the Eurasian Natural Resources Corporation. It is currently under restructuring and is no longer trading under the CAMEC brand. African mineral reserves rank 1st or 2nd for bauxite, cobalt, diamonds, phosphate rocks, platinum-group metals (PGM), vermiculite, and zirconium. Many other minerals are also present in quantity. The 2005 share of world production from African soil is the following : bauxite 9%; aluminium 5%; chromite 44%; cobalt 57%; copper 5%; gold 21%; iron ore 4%; steel 2%; lead (Pb) 3%; manganese 39%; zinc 2%; cement 4%; natural diamond 46%; graphite 2%; phosphate rock 31%; coal 5%; mineral fuels (including coal) & petroleum 13%; uranium 16%.

This book covers the entire spectrum of mineralogy and consolidates its applications in various fields.

— George Beckett

Chapter 1

Introduction

A mineral is a naturally occurring solid chemical substance formed through biogeochemical processes, having characteristic chemical composition, highly ordered atomic structure, and specific physical properties. By comparison, a rock is an aggregate of minerals and/or mineraloids and does not have a specific chemical composition. Minerals range in composition from pure elements and simple salts to very complex silicates with thousands of known forms. The study of minerals is called mineralogy.

Mineral Definition and Classification

To be classified as a true mineral, a substance must be a solid and have a crystalline structure. It must also be a naturally occurring, homogeneous substance with a defined chemical composition.

The International Mineralogical Association approved the following definition in 1995:

> *"A mineral is an element or chemical compound that is normally crystalline and that has been formed as a result of geological processes."*

According to this definition and classification scheme, biogenic materials were excluded from the mineral kingdom:

"Biogenic substances are chemical compounds produced entirely by biological processes without a geological component (e.g., urinary calculi, oxalate crystals in plant tissues, shells of marine molluscs, etc.) and are not regarded as minerals. However, if geological processes were involved in the genesis of the compound, then the product can be accepted as a mineral."

However, other researchers do not adhere to this exclusion rule. Lowenstam (1981), for example, states the following:

> *"Organisms are capable of forming a diverse array of minerals, some of which cannot be formed inorganically in the biosphere."*

The distinction is a matter of classification and less to do with the constituents of the minerals themselves. Skinner (2005) views all solids as potential minerals and includes biominerals in the mineral kingdom, which are those that are created by the metabolic activities of organisms. Inclusion of these biogenic minerals requires a expanded definition of a mineral as:

> *"An element or compound, amorphous or crystalline, formed through* biogeochemical *processes."*

Mineral classification schemes and their definitions are evolving to match recent advances in mineral science. More recent classifications, for example, include an *organic class* – in both the new Dana and the Strunz classification schemes. The organic class includes a very rare group of minerals with hydrocarbons.

The IMA Commission on New Minerals and Mineral Names recently adopted (in 2009) a hierarchical scheme for the naming and classification of mineral groups and group names and established seven commissions and four working groups to review and classify minerals into an official listing of their published names. According to these new rules, "mineral species can be grouped in a number of different ways, on the basis of chemistry, crystal structure, occurrence, association, genetic history, or resource, for example, depending on the purpose to be served by the classification."

Recent advances in high-resolution genetic and x-ray absorption spectroscopy is opening new revelations on the biogeochemical relations between microrganisms and minerals that may make Nickel's (1995) biogenic mineral exclusion obsolete and Skinner's (2005) biogenic mineral inclusion a necessity.

For example, the IMA commissioned 'Environmental Mineralogy and Geochemistry Working Group' deals with minerals in the hydrosphere, atmosphere, and biosphere. Mineral forming microorganisms inhabit the areas that this working group deals with. These organisms exist on nearly every rock, soil, and particle surface spanning the globe reaching depths at 1600 metres below the sea floor (possibly further) and 70 kilometres into the stratosphere (possibly entering the mesosphere). Biologists and geologists have recently started to research and appreciate the magnitude of mineral geoengineering that these creatures are capable of. Bacteria have contributed to the formation of minerals for billions of years and critically define the biogeochemical cycles on this planet. Microorganisms can precipitate metals from solution contributing to the formation of ore deposits in addition to their ability to catalyse mineral dissolution, to respire, precipitate, and form minerals.

Prior to the International Mineralogical Association's listing, over 60 biominerals had been discovered, named, and published. These minerals (a sub-set tabulated in Lowenstam (1981)) are considered minerals proper according to the Skinner (2005) definition. These biominerals are not listed in the International Mineral Association official list of mineral names, however, many of these biomineral representatives are distributed amongst the 78 mineral classes listed in the 'Dana' classification scheme. Another rare class of minerals (primarily biological in origin) include the mineral liquid crystals that are crystalline and liquid at the same time. To date over 80,000 liquid crystaline compounds have been identified.

Concerning the use of the term "mineral" to name this family of liquid crystals, one can argue that the term inorganic would be more appropriate. However, inorganic liquid crystals have long been used for organometallic liquid crystals. Therefore in order to avoid any confusion between these fairly chemically different families, and taking into account that a large number of these liquid crystals occur naturally in nature, we think that the use of the old fashioned but adequate "mineral" adjective taken sensus largo is more specific that an alternative such as "purely inorganic", to name this subclass of the inorganic liquid crystals family.

The Skinner (2005) definition of a mineral takes this matter into account by stating that a mineral can be crystalline *or* amorphous. Liquid mineral crystals are amorphous. Biominerals and liquid mineral crystals, however, are not the primary form of minerals, most are geological in origin, but these groups do help to identify at the margins of what constitutes a mineral proper.

Crystal Structure

A *crystal structure* is the orderly geometric spatial arrangement of atoms in the internal structure of a mineral. There are 14 basic crystal lattice arrangements of atoms in three dimensions, and these are referred to as the 14 "Bravais lattices". Each of these lattices can be classified into one of the seven crystal systems, and all crystal structures currently recognised fit in one Bravais lattice and one crystal system. This crystal structure is based on regular internal atomic or ionic arrangement that is often expressed in the geometric form that the crystal takes. Even when the mineral grains are too small to see or are irregularly shaped, the underlying crystal structure is always periodic and can be determined by X-ray diffraction. Chemistry and crystal structure together define a mineral. In fact, two or more minerals may have the same chemical composition, but differ in crystal structure (these are known as *polymorphs*). For example, pyrite and marcasite are both iron sulfide, but their arrangement of atoms differs. Similarly, some minerals have different chemical compositions, but the same crystal structure: for example, halite (made from sodium and chlorine), galena (made from lead and sulfur) and periclase (made from magnesium and oxygen) all share the same cubic crystal structure.

Crystal structure greatly influences a mineral's physical properties. For example, though diamond and graphite have the same composition (both are pure carbon), graphite is very soft, while diamond is the hardest of all known minerals. This happens because the carbon atoms in graphite are arranged into sheets which can slide easily past each other, while the carbon atoms in diamond form a strong, interlocking three-dimensional network.

There are currently more than 4,000 known minerals, according to the International Mineralogical Association (IMA), which is responsible for the approval of and naming of new mineral species found in nature. Of these, perhaps 100 can be called "common", 50 are "occasional", and the rest are "rare" to "extremely rare".

Mineral Groups and Solid Solution

The chemical composition may vary between end members of a mineral system. For example the plagioclase feldspars comprise a continuous series from sodium and silicon-rich albite ($NaAlSi_3O_8$) to calcium and aluminium-rich anorthite ($CaAl_2Si_2O_8$) with four recognised intermediate compositions between. Mineral-like substances that don't strictly meet the definition are sometimes classified as mineraloids.

Minerals with the same structure and forming solid solutions are named isomorphs, and form series; for example: forsterite and fayalite of the olivine series, ferberite and hubnerite of the wolframite series. Minerals with the same structure and not forming solid solutions are named isotypes, and form groups [classification of minerals (non silicates)]. Minerals with a similar structure are grouped in homeotype families: amphibole and pyroxene families [classification of minerals (silicates)].

Some ion groups with a similar radius can occupy the same structural site in the crystal cell:

- O^{2-} and OH^- with 1.32 and 1.33 Å respectively.
- Si^{4+} and Al^{3+} with 0.42 and 0.51 Å respectively, the charge is neutralised through an exchange of the other cations:
  - o Si^{4+} – (Al^{3+} and Na^+) or (Si^{4+} and Na^+) – (Al^{3+} and Ca^{2+}).
- Larger molecules may have an unoccupied structural site by occupying another unoccupied structural site or by using a divalent cation instead of two monovalent cations, for instance (amphibole family).
- By the end of the 18th century, the minerals were getting chemical formulas. There were some difficulties, as elements were being discovered and isolated for the first time. The minerals: gadolinite-(Y) (first publication: 1802, 09.AJ.20), aeschynite-(Ce) (first publication: 1830, 04.DF.05), vanadinite (first publication: 1838, 08.BN.05), aenigmatite (first publication: 1865, 09.DH.40), labyrinthite (IMA 2002-065, 09.CO.10), illustrate these difficulties.

More recent definitions:

- "A mineral group consists of two or more minerals with the same (isotypic) or essentially the same (homeotypic) structure, and composed of chemically similar elements" (IMA-CNMNC).
- "two structures are considered homeotypic if all essential features of topology are preserved between them" (IUCr).

Minerals and Rocks

A mineral is a naturally occurring solid with a definite chemical composition and a specific crystalline structure. A rock is an aggregate of one or more minerals. (A rock may also include organic remains and mineraloids.) Some rocks are predominantly composed of just one mineral. For example, limestone is a sedimentary rock composed almost entirely of the mineral calcite. Other rocks contain many minerals, and the specific minerals in a rock can vary widely. Some minerals, like quartz, mica or feldspar are common, while others have been found in only four or five locations worldwide. The vast majority of the rocks of the Earth's crust consist of quartz, feldspar, mica, chlorite, kaolin, calcite, epidote, olivine, augite, hornblende, magnetite, hematite, limonite and a few other minerals. Over half of the mineral species known are so rare that they have only been found in a handful of samples, and many are known from only one or two small grains.

Commercially valuable minerals and rocks are referred to as industrial minerals. Rocks from which minerals are mined for economic purposes are referred to as ores (the rocks and minerals that remain, after the desired mineral has been separated from the ore, are referred to as tailings).

Mineral Composition of Rocks

A main determining factor in the formation of minerals in a rock mass is the chemical composition of the mass, for a certain mineral can be formed only when the necessary elements are present in the rock. Calcite is most common in limestones, as these consist essentially of calcium carbonate; quartz is common in sandstones and in certain igneous rocks which contain a high percentage of silica.

Other factors are of equal importance in determining the natural association or paragenesis of rock-forming minerals, principally the mode of origin of the rock and the stages through which it has passed in attaining its present condition. Two rock masses may have very much the same bulk composition and yet consist of entirely different assemblages of minerals.

The tendency is always for those compounds to be formed which are stable under the conditions under which the rock mass originated. A granite arises by the consolidation of a molten magma at high temperatures and great pressures and its component minerals are those stable under such conditions. Exposed to moisture, carbonic acid and other subaerial agents at the ordinary temperatures of the Earth's surface, some of these original minerals, such as quartz and white mica are relatively stable and remain unaffected; others weather or decay and are replaced by new combinations. The feldspar passes into kaolinite, muscovite and quartz, and any mafic minerals such as pyroxenes, amphiboles or biotite have been present they are often altered to chlorite, epidote, rutile and other substances. These changes are accompanied by disintegration, and the rock falls into a loose, incoherent, earthy mass which may be regarded as a sand or soil. The materials thus formed may be washed away and deposited as sandstone or siltstone.

The structure of the original rock is now replaced by a new one; the mineralogical constitution is profoundly altered; but the bulk chemical composition may not be very different. The sedimentary rock may again undergo metamorphism.

If penetrated by igneous rocks it may be recrystallised or, if subjected to enormous pressures with heat and movement during mountain building, it may be converted into a gneiss not very different in mineralogical composition though radically different in structure to the granite which was its original state.

Physical Properties of Minerals

Classifying minerals can range from simple to very difficult. A mineral can be identified by several physical properties, some of them being sufficient for full identification without equivocation. In other cases, minerals can only be classified by more complex optical, chemical or X-ray diffraction analysis; these methods, however, can be costly and time-consuming.

Physical properties commonly used are:

- *Crystal structure and habit*: A mineral may show good crystal habit or form, or it may be massive, granular or compact with only microscopically visible crystals.

Figure 1: Talc

Figure 2: Rough diamond.

- *Hardness*: the physical hardness of a mineral is usually measured according to the Mohs scale. This scale is relative and goes from 1 to 10. Minerals with a given Mohs hardness can scratch the surface of any mineral that has a lower hardness than itself.
  - o Mohs hardness scale:
    - * Talc $Mg_3Si_4O_{10}(OH)_2$
    - * Gypsum $CaSO_4 \cdot 2H_2O$
    - * Calcite $CaCO_3$
    - * Fluorite CaF_2
    - * Apatite $Ca_5(PO_4)_3(OH,Cl,F)$
    - * Orthoclase $KAlSi_3O_8$
    - * Quartz SiO_2
    - * Topaz $Al_2SiO_4(OH,F)_2$
    - * Corundum Al_2O_3
    - * Diamond C (pure carbon).
- *Luster* indicates the way a mineral's surface interacts with light and can range from dull to glassy (vitreous).
  - o Metallic – high reflectivity like metal: galena and pyrite
  - o Sub-metallic – slightly less than metallic reflectivity: magnetite
  - o Non-metallic lusters:

  * *Adamantine* – brilliant, the luster of diamond also cerussite and anglesite
  * *Vitreous* – the luster of a broken glass: quartz
  * *Pearly* – iridescent and pearl-like: talc and apophyllite
  * *Resinous* – the luster of resin: sphalerite and sulfur
  * *Silky* – a soft light shown by fibrous materials: gypsum and chrysotile
  * Dull/earthy – shown by finely crystallised minerals: the kidney ore variety of hematite

- *Diaphaneity* describes how well light passes through a mineral; there are three basic degrees of transparency:
  - o *Transparent* objects can be seen through a transparent mineral, such as a clear quartz crystal
  - o *Translucent* light passes through the mineral but no objects can be seen
  - o *Opaque* no light passes through the mineral.

Many minerals range from transparent to translucent or translucent to opaque. Calcite, for instance, can be translucent or opaque. Some minerals that are naturally translucent become opaque with weathering.

- *Colour* indicates the appearance of the mineral in reflected light or transmitted light for translucent minerals (i.e. what it looks like to the naked eye).
  - o *Iridescence* – the *play of colours* due to surface or internal interference. Labradorite exhibits internal iridescence whereas hematite and sphalerite often show the surface effect.
- *Streak* refers to the colour of the powder a mineral leaves after rubbing it on an unglazed porcelain *streak plate*. Note that this is not always the same colour as the original mineral.
- *Cleavage* describes the way a mineral may split apart along various planes. In thin sections, cleavage is visible as thin parallel lines across a mineral.
- *Fracture* describes how a mineral breaks when broken contrary to its natural cleavage planes.
  - o *Chonchoidal fracture* is a smooth curved fracture with concentric ridges of the type shown by glass.

- o *Hackley* is jagged fracture with sharp edges.
- o *Fibrous*
- o *Irregular*

- *Specific gravity* relates the mineral mass to the mass of an equal volume of water, namely the density of the material. While most minerals, including all the common rock-forming minerals, have a specific gravity of 2.5–3.5, a few are noticeably more or less dense, e.g. several sulfide minerals have high specific gravity compared to the common rock-forming minerals.
- Other properties: fluorescence (response to ultraviolet light), magnetism, radioactivity, tenacity (response to mechanical induced changes of shape or form), piezoelectricity and reactivity to dilute acids.

Chemical Properties of Minerals

Minerals may be classified according to chemical composition. They are here categorised by anion group. The list below is in approximate order of their abundance in the Earth's crust. The list follows the Dana classification system which closely parallels the Strunz classification.

Silicate Class

Figure 3: Quartz

The largest group of minerals by far are the silicates (most rocks are e"95% silicates), which are composed largely of silicon and oxygen, with the addition of ions such as aluminium, magnesium, iron, and calcium. Some important rock-forming silicates include the feldspars, quartz, olivines, pyroxenes, amphiboles, garnets, and micas.

Carbonate Class

The carbonate minerals consist of those minerals containing the anion $(CO_3)^{2-}$ and include calcite and aragonite (both calcium carbonate), dolomite (magnesium/calcium carbonate) and siderite (iron carbonate). Carbonates are commonly deposited in marine settings when the shells of dead planktonic life settle and accumulate on the sea floor. Carbonates are also found in evaporitic settings (e.g. the Great Salt Lake, Utah) and also in karst regions, where the dissolution and reprecipitation of carbonates leads to the formation of caves, stalactites and stalagmites. The carbonate class also includes the nitrate and borate minerals.

Sulfate Class

Figure 4: Hanksite, $Na_{22}K(SO_4)_9(CO_3)_2Cl$, one of the few minerals that is considered a carbonate and a sulfate

Sulfate minerals all contain the sulfate anion, SO_4^{2-}. Sulfates commonly form in evaporitic settings where highly saline waters slowly evaporate, allowing the formation of both sulfates and halides at the water-sediment interface. Sulfates also occur in hydrothermal vein systems as gangue minerals along with sulfide ore minerals. Another occurrence is as secondary oxidation products of original sulfide minerals. Common sulfates include anhydrite (calcium sulfate), celestine (strontium sulfate), barite (barium sulfate), and gypsum (hydrated calcium sulfate). The sulfate class also includes the chromate, molybdate, selenate, sulfite, tellurate, and tungstate minerals.

Halide Class

Figure 5: Halite

The halide minerals are the group of minerals forming the natural salts and include fluorite (calcium fluoride), halite (sodium chloride), sylvite (potassium chloride), and sal ammoniac (ammonium chloride). Halides, like sulfates, are commonly found in evaporite settings such as salt lakes and landlocked seas such as the Dead Sea and Great Salt Lake. The halide class includes the fluoride, chloride, bromide and iodide minerals.

Oxide Class

Oxide minerals are extremely important in mining as they form many of the ores from which valuable metals can be extracted. They also carry the best record of changes in the Earth's magnetic field. They commonly occur as precipitates close to the Earth's surface, oxidation products of other minerals in the near surface weathering zone, and as accessory minerals in igneous rocks of the crust and mantle.

Common oxides include hematite (iron oxide), magnetite (iron oxide), chromite (iron chromium oxide), spinel (magnesium aluminium oxide – a common component of the mantle), ilmenite (iron titanium oxide), rutile (titanium dioxide), and ice (hydrogen oxide). The oxide class includes the oxide and the hydroxide minerals.

Sulfide Class

Many sulfide minerals are economically important as metal ores. Common sulfides include pyrite (iron sulfide – commonly known as *fools' gold*), chalcopyrite (copper iron sulfide), pentlandite (nickel iron sulfide), and galena (lead sulfide). The sulfide class also includes the selenides, the tellurides, the arsenides, the antimonides, the bismuthinides, and the sulfosalts (sulfur and a second anion such as arsenic).

Phosphate Class

The phosphate mineral group actually includes any mineral with a tetrahedral unit AO_4 where A can be phosphorus, antimony, arsenic or vanadium. By far the most common phosphate is apatite which is an important biological mineral found in teeth and bones of many animals. The phosphate class includes the phosphate, arsenate, vanadate, and antimonate minerals.

Element Class

The elemental group includes native metals and intermetallic elements (gold, silver, copper), semi-metals and non-metals (antimony, bismuth, graphite, sulfur). This group also includes natural alloys, such as electrum (a natural alloy of gold and silver), phosphides, silicides, nitrides and carbides (which are usually only found naturally in a few rare meteorites).

Organic Class

The organic mineral class includes biogenic substances in which geological processes have been a part of the genesis or origin of the existing compound. Minerals of the organic class include various oxalates, mellitates, citrates, cyanates, acetates, formates, hydrocarbons and other miscellaneous species. Examples include whewellite, moolooite, mellite, fichtelite, carpathite, evenkite and abelsonite.

Bowen's Reaction Series

Within the field of geology, Bowen's reaction series is the work of the petrologist, Norman L. Bowen who was able to explain why certain types of minerals tend to be found together while others are almost never associated with one another. He experimented in the early 1900s with powdered rock material that was heated until it melted and then allowed to cool to a target temperature whereupon he observed the types of minerals that formed in the rocks produced. He repeated this process with progressively cooler temperatures and the results he obtained led him to formulate his reaction series which is still accepted today as the idealised progression of minerals produced by cooling magma. Based upon Bowen's work, one can infer from the minerals present in a rock the relative conditions under which the material had formed.

Description

Figure 6: Olivine weathering to iddingsite within a mantle xenolith, demonstrating the principles of the Goldich dissolution series

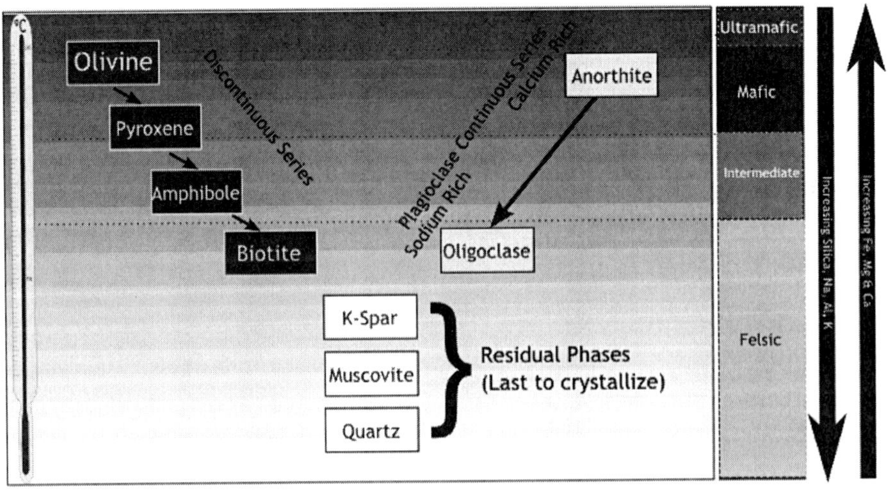

Figure 7: Bowen's Reaction Series

The series is broken into two branches, the continuous and the discontinuous. The branch on the right is the continuous. The minerals at the top of the illustration (given aside) are first to crystallise and so the temperature gradient can be read to be from high to low with the high temperature minerals being on the top and the low temperature ones on the bottom.

Since the surface of the Earth is a low temperature environment compared to the zones of rock formation, the chart also easily shows the stability of minerals with the ones at bottom being most stable and the ones at top being quickest to weather, known as the Goldich dissolution series.

This is because minerals are most stable in the conditions closest to those under which they had formed. Put simply, the high temperature minerals, the first ones to crystallise in a mass of magma, are most unstable at the Earth's surface and quickest to weather because the surface is most different from the conditions under which they were created while the low temperature minerals are much more stable because the conditions at the surface are much more similar to the conditions under which they formed.

Dietary Mineral

Dietary minerals (also known as mineral nutrients) are the chemical elements required by living organisms, other than the four elements carbon, hydrogen, nitrogen, and oxygen present in common organic molecules. Examples of mineral elements include calcium, magnesium, potassium, sodium, zinc, and iodine. Most minerals that enter into the dietary physiology of organisms consist of simple chemical elements. Larger aggregates of minerals need to be broken down for absorption. Bacteria play an essential role in the weathering of primary minerals that results in the release of nutrients for their own nutrition and for the nutrition of others in the ecological food chain. Scientists are only recently starting to appreciate the magnitude and role that microorganisms have in the global cycling and formation of biological minerals. Plants absorb dissolved minerals in soils, which are subsequently picked up by the herbivores that eat them and so on, the minerals move up the food chain. Larger organisms may also consume soil (geophagia) and visit mineral licks to obtain limiting mineral nutrients they are unable to acquire through other components of their diet.

Essential Chemical Elements

Some sources state that sixteen chemical elements are *required* to support human biochemical processes by serving structural and functional roles as well as electrolytes: As many as 26 elements are suggested to be used by mammals, as a result of studies of biochemical, special uptake, and metabolic handling studies. However, many of these additional elements have no well-defined biochemical function known at present. Most of the known and suggested dietary elements are of relatively low atomic weight, and are reasonably common on land, or at least, common in the ocean (iodine, sodium):

Dietary Nutrition

Dietitians may recommend that dietary elements are best supplied by ingesting specific foods rich with the chemical element(s) of interest. The elements may be naturally present in the food (e.g., calcium in dairy milk) or added to the food (e.g., orange juice fortified with calcium; iodized salt, salt fortified with iodine). Dietary supplements can be formulated to contain several different chemical elements (as compounds), a combination of vitamins and/or other chemical compounds, or a single element (as a compound or mixture of compounds), such as calcium (as carbonate, citrate, etc.) or magnesium (as oxide, etc.), chromium (usually as picolinate) or iron (as bis-glycinate).

The dietary focus on chemical elements derives from an interest in supporting the biochemical reactions of metabolism with the required elemental components. Appropriate intake levels of certain chemical elements have been demonstrated to be required to maintain optimal health. Diet can meet all the body's chemical element requirements, although supplements can be used when some requirements (e.g., calcium, which is found mainly in dairy products) are not adequately met by the diet, or when chronic or acute deficiencies arise from pathology, injury, etc. Research has supported that altering inorganic mineral compounds (carbonates, oxides, etc.) by reacting them with organic ligands (amino acids, organic acids, etc.) improves the bioavailability of the supplemented mineral.

Other Elements

Many elements have been suggested as essential, but such claims have usually not been confirmed. Definitive evidence for efficacy comes from the characterisation of a biomolecule containing the element with an identifiable and testable function. One problem with identifying efficacy is that some elements are innocuous at low concentrations and are pervasive (examples: silicon and nickel in soild and dust), so proof of efficacy is lacking because deficiencies are difficult to reproduce.

Element	*Description*	*Excess*
Sulfur	Relatively large quantities of sulfur are required, but there is no RDA, as the sulfur is obtained from and used for amino acids, and therefore should be adequate in any diet containing enough protein.	(primarily associated with compounds)
Cobalt	Cobalt is required in the synthesis of vitamin B_{12}, but because bacteria are required to synthesize the vitamin, it is usually considered part of vitamin B_{12} deficiency rather than its own dietary element deficiency.	Cobalt poisoning
Nickel	There have been occasional studies asserting the essentiality of nickel, but it currently has no RDA.	Nickel toxicity
Chromium	Chromium has been described as nonessential to mammals, with some role in sugar metabolism in humans. Despite a market for the supplement chromium picolinate, definitive biochemical evidence for a physiological function is lacking.	Chromium toxicity Fluoride poisoning
Fluorine	Fluorine (as Fluoride) is not generally considered an essential mineral element because humans do not require it for growth or to sustain life. However, if one considers the prevention of dental caries an important criterion in determining essentiality, then fluoride might well be considered an essential trace element. However, recent research indicates that the primary action of fluoride occurs topically (at the surface).	
Boron	Boron has been found to be essential for the utilisation of vitamin D and calcium in the body.	

Strontium	Strontium has been found to be involved in the utilisation of calcium in the body. It has promoting action on calcium uptake into bone at moderate dietary strontium levels, but a rachitogenic (rickets-producing) action at higher dietary levels.	Rachitogenic
Other	Arsenic, silicon, and vanadium have established, albeit specialised, biochemical roles as structural or functional cofactors in other organisms, and are possibly, even probably, used by mammals (including humans). By contrast, tungsten, bromine, and cadmium have specialised biochemical uses in certain lower organisms, but these elements appear not to be utilised by humans.	Multiple

Mineral Ecology

Recent studies have shown a tight linkage between living organisms and minerals on this planet. This has let to the redefinition of minerals as "an element or compound, amorphous or crystalline, formed through 'biogeochemical' processes. The addition of `bio' reflects a greater appreciation, although an incomplete understanding, of the processes of mineral formation by living forms." Biologists and geologists have only recently started to appreciate the magnitude of mineral biogeoengineering. Bacteria have contributed to the formation of minerals for billions of years and critically define the biogeochemical mineral cycles on this planet. Microorganisms can precipitate metals from solution contributing to the formation of ore deposits in addition to their ability to catalyse mineral dissolution, to respire, precipitate, and form minerals.

Most minerals are inorganic in nature. Mineral nutrients refers to the smaller class of minerals that are metabolised for growth, development, and vitality of living organisms. Mineral nutrients are recycled by bacteria that are freely suspended in the vast water columns of the worlds oceans. They absorb dissolved organic matter containing mineral nutrients as they scavenge through the dying individuals that fall out of large phytoplankton blooms. Flagellates are effective bacteriovores and are also commonly found in the marine water column. The flagellates are preyed upon by zooplankton while the phytoplankton concentrates on the larger particulate matter that is suspended in the water column as they are consumed by larger zooplankton, with fish as the top predator. Mineral nutrients cycle through this marine food chain, from bacteria and phytoplankton to flagellates and zooplankton who are then eaten by fish. The bacteria are important in this chain because only they have the physiological ability to absorb the dissolved mineral nutrients from the sea. These recycling principals from marine environments apply to many soil and freshwater ecosystems as well.

Goldich Dissolution Series

The Goldich dissolution series is a way of predicting the relative stability or weathering rate of various minerals on the Earth's surface. S. S. Goldich came up with the series in 1938 after studying soil profiles. He found that minerals that form at higher temperatures and pressures are less stable on the surface than minerals that form at lower temperatures and pressures. This pattern follows the same pattern of the Bowen's reaction series, with the minerals that are first to crystallise also the first the undergo chemical weathering.

Chapter 2

Types of Minerals

Abelsonite

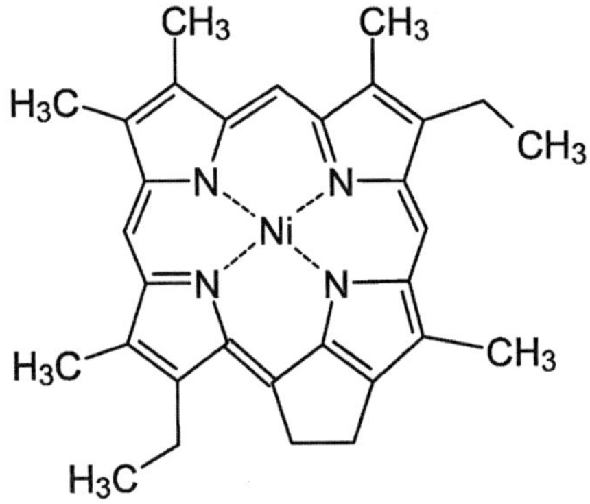

Figure 1: Abelsonite

Abelsonite (synonymous with nickel porphyrin), is an organic porphyrin mineral first described in 1975 for specimens in oil shale from the Green River Formation in eastern Uintah County, Utah. It is named after Philip Hauge Abelson, an American physicist.

Abenakiite-(Ce)

Abenakiite-(Ce) (or IMA1991-054) is a mineral of sodium, cerium, neodymium, lanthanum, praseodymium, thorium, samarium, oxygen, sulfur, carbon, phosphorus, and silicon with a chemical formula $Na_{26}(Ce,Nd,La,Pr,Th,Sm)_6$-$[O\,|\,SO_2\,|\,(CO_3)_6\,|\,(PO_4)_6\,|\,Si_6O_{18}]$. It is named after the Abenaki, an Algonquian Indian tribe of New England. Its Mohs scale rating is 4 to 5.

Abernathyite

Abernathyite is a rare hydrous uranium oxide mineral with chemical formula $K_2(UO_2)_2(AsO_4)_2 \cdot 6H_2O$. It occurs as a secondary fracture filling mineral. It crystallises in the tetragonal system and forms yellow vitreous tabular crystals with perfect cleavage. It has a specific gravity of 3.3 and a hardness of 2.5.

It was first reported from a Colourado Plateau-type uranium deposit near Moab, Utah. It is named after mine operator Jess Abernathy.

Abhurite

Abhurite is a mineral of tin, oxygen, hydrogen, and chlorine with the formula $Sn_3O(OH)_2Cl_2$. It is named after Sharm Abhur, a cove of the Red Sea. Its Mohs scale rating is 2.

Abswurmbachite

Abswurmbachite is a copper manganese silicate mineral $((Cu,Mn^{2+})Mn^{3+}{}_6O_8SiO_4)$. It was first described in 1991 and named after Irmgard Abs-Wurmbach, a German mineralogist. It crystallises in the tetragonal system. Its Mohs scale rating is 6.5 and a specific gravity of 4.96. It has a metallic luster and its colour is jet black, with light brown streaks

Acanthite

Figure 2: Acanthite on calcite - Locality: Freiberg District, Erzgebirge, Saxony, Germany- Scale is one inch with a rule at one cm

Acanthite, Ag_2S, crystallises in the monoclinic system and is the stable form of silver sulfide below 173 °C. Argentite is the stable form above that temperature. As argentite cools below that temperature its cubic form is distorted to the monoclinic form of acanthite. Below 173 °C acanthite forms directly. Acanthite is the only stable form in normal air temperature.

Occurrence

Figure 3: Acanthite - Locality: Chispas Mine, Arizpe, Mun. de Arizpe, Sonora, Mexico - Scale is one inch with a rule at one cm

Acanthite is a common silver mineral in moderately low-temperature hydrothermal veins and in zones of supergene enrichment. It occurs in association with native silver, pyrargyrite, proustite, polybasite, stephanite, aguilarite, galena, chalcopyrite, sphalerite, calcite and quartz.

Acanthite was first described in 1855 for an occurrence in the Jáchymov (St Joachimsthal) District, Krušné Hory Mts (Erzgebirge), Karlovy Vary Region, Bohemia, Czech Republic. The name is from the Greek "akantha" meaning thorn or arrow, in reference to its crystal shape.

Actinolite

Figure 4: Actinolite crystal from Portugal

Actinolite is an amphibole silicate mineral with the chemical formula $Ca_2(Mg,Fe)_5Si_8O_{22}(OH)_2$.

Etymology

The name *actinolite* is derived from the Greek word *aktis*, meaning "beam" or "ray", because of the mineral's fibrous nature. (This word is also the origin of the name of the chemical element actinium.)

Mineralogy

Actinolite is an intermediate member in a solid-solution series between magnesium-rich tremolite, $Ca_2Mg_5Si_8O_{22}(OH)_2$, and iron-rich ferro actinolite, $Ca_2Fe_5Si_8O_{22}(OH)_2$. Mg and Fe ions can be freely exchanged in the crystal structure. Like tremolite, asbestiform actinolite is regulated as asbestos.

Occurrence

Actinolite is commonly found in metamorphic rocks, such as contact aureoles surrounding cooled intrusive igneous rocks. It also occurs as a product of metamorphism of magnesium-rich limestones.

The old mineral name *uralite* is at times applied to an alteration product of primary pyroxene by a mixture composed largely of actinolite. The metamorphosed gabbro or diabase rock bodies, referred to as epidiorite, contain a considerable amount of this *uralitic* alteration.

Some forms of asbestos are formed from fibrous actinolite, the fibres being so small that they can enter the lungs and damage the alveoli.

Gemology

Some forms of actinolite are used as gemstones. One is nephrite, one of the two types of jade (the other being jadeite, a variety of pyroxene).

Another gem variety is the chatoyant form known as *cat's-eye actinolite*. This stone is translucent to opaque, and green to yellowish green colour. This variety has had the misnomer *jade cat's-eye*. Transparent actinolite is rare and is faceted for gem collectors. Major sources for these forms of actinolite are Taiwan and Canada. Other sources are Madagascar, Tanzania, and the United States.

Acuminite

Acuminite is a rare halide mineral of with chemical formula: $SrAlF_4(OH)\cdot(H_2O)$. Its name comes from the Latin word *acumen*, meaning "spear point". Its Mohs scale rating is 3.5.

Acumenite has only been described from its type locality of the cryolite deposit in Ivigtut, Greenland.

Adamite

Figure 5: Yellow-green adamite in limonite

Adamite is a zinc arsenate hydroxide mineral, Zn_2AsO_4OH. It is a mineral that typically occurs in the oxidised or weathered zone above zinc ore occurrences. Pure adamite is colourless, but usually it possess yellow colour due to Fe compounds admixture. Tints of green also occur and are connected with copper substitutions in the mineral structure. Olivenite is a copper arsenate that is isostructural with adamite and there is considerable substitution between zinc and copper resulting in an intermediate called *cuproadamite*. Zincolivenite is a recently discovered mineral being an intermediate mineral with formula $CuZn(AsO_4)(OH)$. Manganese, cobalt, and nickel also substitute in the structure. An analogous zinc phosphate, *tarbuttite*, is known.

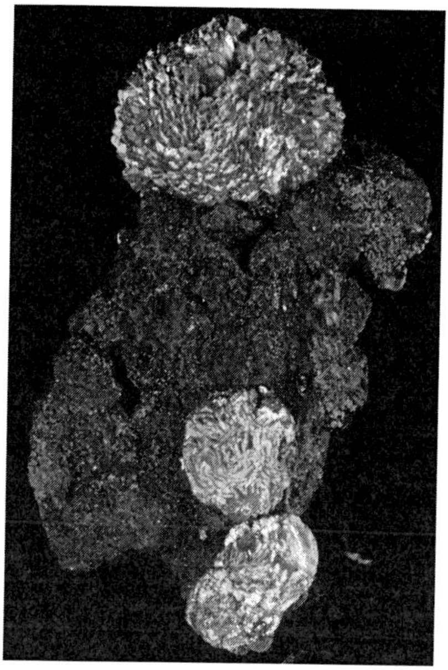

Figure 6: Adamite on limonite from the Gold Hill District Tooele County, Utah, USA - Scale at bottom is approx. 2.5 cm.

Occurrence

Adamite occurs as a secondary mineral in the oxidised zone of zinc- and arsenic-bearing hydrothermal mineral deposits. It occurs in association with smithsonite, hemimorphite, scorodite, olivenite, calcite, quartz and iron and manganese oxides.

The yellow to bright lime-green coloured crystals and druze along with its distinctive fluorescence make adamite a favourite among mineral collectors. Found in Mapimi, Mexico; Greece; and California and Utah in the United States.

Adamite was named after the French mineralogist Gilbert-Joseph Adam (1795-1881). It was first described in 1866 for an occurrence at the type locality of Chañarcillo, Copiapó Province, Atacama Region, Chile.

Adamsite-(Y)

Adamsite-(Y) (previously IMA1999-020) $NaY(CO_3)_2 \cdot 6H_2O$ is a mineral of sodium, yttrium, carbon, oxygen, and hydrogen. It is named after Prof. Frank Dawson Adams, a Canadian geologist. Its Mohs scale rating is 3.

Adelite

The rare mineral adelite, is a calcium, magnesium, arsenate with chemical formula $CaMgAsO_4OH$. It forms a solid solution series with the vanadium bearing mineral gottlobite. Various transition metals substitute for magnesium and lead replaces calcium leading to a variety of similar minerals in the *adelite - duftite group.*

Adelite forms variably coloured (blue, green, yellow and grey) crystals in the orthorhombic crystal system. The form is typically massive. It has a Mohs hardness rating of 5 and a specific gravity of 3.73 to 3.79.

It was first described in 1891 from Värmland, Sweden. Its name comes from the Greek word for *indistinct.*

Aegirine

Figure 7: Aegirine crystals in matrix

Aegirine is a member of the clinopyroxene group of inosilicates. Aegirine is the sodium endmember of the aegirine-augite series. Aegirine has the chemical formula $NaFeSi_2O_6$ in which the iron is present as Fe^{3+}. In the aegirine-augite series the sodium is variably replaced by calcium with iron(II) and magnesium replacing the iron(III) to balance the charge. Aluminium also substitutes for the iron(III). It is also known as *acmite*, which is a fibrous, green-coloured variety.

Aegirine occurs as dark green monoclinic prismatic crystals. It has a glassy luster and perfect cleavage. The Mohs hardness varies from 5 to 6 and the specific gravity is 3.2 to 3.4.

Figure 8: Aegirine

Commonly occurs in alkalic igneous rocks, nepheline syenites, carbonatites and pegmatites. Also in regionally metamorphosed schists, gneisses, and iron formations; in blueschist facies rocks, and from sodium metasomatism in granulites. It may occur as an authigenic mineral in shales and marls. It occurs in association with potassic feldspar, nepheline, riebeckite, arfvedsonite, aenigmatite, astrophyllite, catapleiite, eudialyte, serandite and apophyllite.

Localities include Mont Saint-Hilaire, Quebec, Canada; Kongsberg, Norway; Narsarssuk, Greenland; Kola Peninsula, Russia; Magnet Cove, Arkansas, USA; Kenya; Scotland and Nigeria.

It was first described in 1835 for an occurrence in Rundemyr, Øvre Eiker, Buskerud, Norway. Aegirine was named after Ægir, the

Teutonic god of the sea. A synonym for the mineral is *acmite* in reference to the typical pointed crystals.

Aenigmatite

Aenigmatite, also known as *Cossyrite* after Cossyra, the ancient name of Pantelleria, is a sodium, iron, titanium inosilicate mineral. The chemical formula is $Na_2Fe^{2+}{}_5TiSi_6O_{20}$ and its structure consists of single tetrahedral chains with a repeat unit of four and complex side branches. It forms brown to black triclinic lamellar crystals. It has Mohs hardness of 5.5 to 6 and specific gravity of 3.74 to 3.85. Aenigmatite forms a solid-solution series with wilkinsonite, $Na_2Fe^{2+}{}_4Fe^{3+}{}_2Si_6O_{20}$.

Aenigmatite is primarily found in peralkaline volcanic rocks, pegmatites, and granites as well as silica-poor intrusive rocks. It was first described by August Breithaupt in 1865 for an occurrence in the Ilimaussaq intrusive complex of southwest Greenland.

It was also reported from the Kaidun meteorite, possibly a Mars meteorite, which landed on March 1980 in South Yemen. Other notable studied occurrences include:

- Narsaarsuk and elsewhere in Greenland.
- The Khibiny and Lovozero alkaline massifs on Kola Peninsula, Russia.
- The volcanic island of Pantelleria, Italy.
- In the USA, from Granite Mountain, near Little Rock, Pulaski County, Arkansas, and Santa Rosa, Sonoma County, California.
- In Australia, from Warrumbungle volcano, Nandewar volcano, and the Mount Warning complex, New South Wales; and the Peak Range Province, Queensland.
- In Canada, from Mount Edziza, the Ilgachuz and Rainbow Range shield complexes.
- From Logan Point quarry, Dunedin volcano, New Zealand.

Aerinite

Aerinite ($Ca_4(Al,Fe,Mg)_{10}Si_{12}O_{35}(OH)_{12}CO_3 \cdot 12H_2O$) is a bluish-purple inosilicate mineral. It crystallises in the monoclinic system and occurs as fibrous masses and coatings. It has a dark, vitreous luster, a specific gravity of 2.48 and a Mohs hardness of 3.

Figure 9: Aerinite from Spain

It is a low-temperature hydrothermal phase occurring in zeolite facies alteration of mafic rocks. Associated minerals include prehnite, scolecite and mesolite.

Its name comes from a Greek root "aerinos," meaning "atmosphere" or "sky".

Aeschynite-(Ce)

Aeschynite-(Ce) (or Aschynite, Eschinite, Eschynite) is a rare earth mineral of cerium, calcium, iron, thorium, titanium, niobium, oxygen, and hydrogen with formula: $(Ce,Ca,Fe,Th)(Ti,Nb)_2(O,OH)_6$. Its name comes from the Greek word for "shame". The "-(Ce)" means it has more cerium than the yttrium variety aeschynite-(Y). Its Mohs scale rating is 5-6.

Aeschynite-(Nd)

Aeschynite-(Nd) is a rare earth mineral of neodymium, cerium, calcium, thorium, titanium, niobium, oxygen, and hydrogen with formula: $(Nd,Ce,Ca,Th)(Ti,Nb)_2(O,OH)_6$. Its name comes from the Greek word for "shame".

It was first reported for an occurrence in Bayan Obo Inner Mongolia in 1982. In that rare earth mining deposit it occurs in veins within metamorphosed dolomite and slate. It occurs associated with aegirine, riebeckite, barite, fluorite, albite, phlogopite and magnetite.

Afghanite

Figure 10: Afghanite

Afghanite, $(Na,Ca,K)_8[Al_6Si_6O_{24}](SO_4,Cl_2,CO_3)_3 \cdot 0.5H_2O$, is a hydrous sodium, calcium, potassium, sulfate, chloride, carbonate alumino-silicate mineral. Afghanite is a feldspathoid of the cancrinite group and typically occurs with sodalite group minerals. It forms blue to colourless, typically massive crystals in the hexagonal crystal system. It has a Mohs hardness of 5.5 to 6 and a specific gravity of 2.55 to 2.65. It has refractive index values of nù=1.523 and nå=1.529. It has one direction of perfect cleavage and exhibits conchoidal fracture. It fluoresces a bright orange.

It was discovered in 1968 in the Lapis-lazuli Mine, Sar-e-Sang, Badakhshan Province, Afghanistan and takes its name from that country. It has also been described from localities in Germany, Italy, the Pamir Mountains, Tajikistan, near Lake Baikal in Siberia, New York and Newfoundland. It occurs as veinlets in lazurite crystals in the Afghan location and in altered limestone xenoliths within pumice in Pitigliano, Tuscany, Italy.

Afwillite

Figure 11: Afwillite from Crestmore Quarry, Riverside County, California

Afwillite is a calcium hydroxide nesosilicate mineral with formula $Ca_3(SiO_3OH)_2.2H_2O$. It occurs as glassy, colourless to white prismatic monoclinic crystals. Its Mohs scale hardness is between 3 and 4. It occurs as an alteration mineral in contact metamorphism of limestone.

It was first described in 1925 for an occurrence in the Dutoitspan Mine, Kimberley, South Africa and was named for Alpheus Fuller Williams (1874-1953), a past official of the De Beers diamond company.

Afwillite is typically found in veins of spurrite and it belongs to the nesosilicate sub-class. It is monoclinic, its space group is P2 and its point group is 2.

Formation of Afwillite

It is suggested that afwillite forms in fractured veins of the mineral spurrite. Jennite, afwillite, oyelite and calcite are all minerals that form in layers within spurrite veins. It appears that the afwillite, as well as the calcite, forms from precipitated fluids. The jennite is actually an alteration of the afwillite, but both formed from calcium silicates through hydration. Laboratory studies determined that afwillite forms at a temperature below 200°C, usually around 100°C. Afwillite and spurrite are formed through contact metamorphism of limestone. Contact metamorphism is caused by the interaction of rock with heat and/or fluids from a nearby crystallising silicate magma.

Structure and Properties

Afwillite has a complex monoclinic structure, and the silicon tetrahedra in the crystal structure are held together by hydrogen bonds. It has perfect cleavage parallel to its (101) and poor cleavage parallel to its (100) faces. It is biaxial and its 2V angle, the measurement from one optical axis to the other optical axis, is 50 – 56 degrees. When viewed under crossed polarisers in a petrographic microscope, it displays first-order orange colours, giving a maximum birefringence of 0.0167 (determined by using the Michel- Levy chart). Afwillite is optically positive. Additionally, it has a prismatic crystal habit. Under a microscope afwillite looks like wollastonite, which is in the same family as afwillite.

Awfillite is composed of double chains that consist of calcium and silicon polyhedral connected to each other by sharing corners and edges. This causes continuous sheets to form parallel to its miller index [-101] faces. The sheets are bonded together by hydrogen bonds and are all connected by Ca-Si-O bonds (Malik and Jeffery 1976). Each calcium atom is in 6-fold octahedral coordination with the oxygen, and the silicon is in 4-fold tetrahedral coordination around the oxygen. Around each silicon there is one OH group and there are three oxygens that neighbour them. The silicon tetrahedra are arranged so that they share an edge with calcium(1), and silicon(2) shares edges with the calcium(2) and calcium(3) polyhedral. The silicon tetrahedra are held together by the OH group and hydrogen bonding occurs between the hydrogen in the OH and the silicon tetrahedra. Hydrogen bonding is caused because the positive ion, hydrogen, is attracted to negatively charge ions which, in this case, are the silicon tetrahedra.

Occurrence in Concrete

Afwillite is one of the calcium silicates that form when Portland cement sets to form concrete. The cement gets its strength from the hydration of its di- and tri- calcium silicates.

Agardite

Figure 12: A rich example of Agardite-(Ce) in the form of pistachio-green acicular crystals on contrasting matrix

Agardite is a mineral group consisting of *agardite-(Ce), agardite-(Nd), agardite-(La), and agardite-(Y).* They comprise a group of minerals that are hydrous hydrated arsenates of rare earth elements and copper, which contain variable amounts of calcium and sometimes lead. Yttrium, cerium, neodymium, lanthanum as well as trace to minor amounts of other rare earth elements are present in their structure.

Agardite-(Y) is probably the most often found representative. The general formula for the group is $(REE,Ca)Cu_6(AsO_4)_3(OH)_6 \cdot 3H_2O$. They form needle-like yellow-green (variably hued) crystals in the hexagonal crystal system.

They were first described in 1970 in the Black Forest, Germany. They were named after Jules Agard, a French geologist.

Agrellite

Agrellite ($NaCa_2Si_4O_{10}F$) is a mineral found in Quebec, Canada. It is also known from few other areas.

Agrellite displays pink fluorescence under both shortwave and longwave ultraviolet light.

It is named in honour of Stuart Olof Agrell.

Agrinierite

Agrinierite ($K_2(Ca,Sr)(UO_2)_3O_3(OH)_2 \cdot 5H_2O$) is a mineral often found in the oxidation zone of uranium deposits. It is named for Henry Agrinier, an engineer for the Commissariat à l'Énergie Atomique.

Aguilarite

Figure 13: Aguilarite - San Carlos Mine, La Luz, Guanajuato municipality, Mexico. Specimen height is 3.9 cm.

Aguilarite (Ag_4SeS) is a sulfosalt mineral found in the San Carlos mine in Mexico. Its crystal system is orthorhombic.

Aheylite

Figure 14: Size: 3.5×2.3×0.8 cm. Translucent ~0.2 cm spheres of aheylite are perched on dark cassiterite, together with elongated quartz crystals.

Aheylite is a rare phosphate mineral with formula $(Fe^{2+}, Zn)Al_6[(OH)_4 | (PO_4)_2]_2 \cdot 4(H_2O)$. It occurs as pale blue to pale green triclinic crystal masses.

Name and Discovery

It was first described for an occurrence in the Huanuni mine, Huanuni, Oruro Department, Bolivia, and named for Allen V. Heyl (1918–2008), an economic geologist for the United States Geological Survey.

Occurrence

In addition to the type locality in Bolivia it has been reported from the Bali Lo prospect in the Capricorn Range, Western Australia and the Les Montmins Mine, Auvergne, France.

It is a turquoise group mineral and occurs as a late hydrothermal phase in a tin deposit associated with variscite, vivianite, wavellite, cassiterite, sphalerite, pyrite and quartz in the type locality.

Ajoite

Figure 15: Ajoite from Arizona (size: 2.4×2.4×2.0 cm)

Ajoite is a hydrated sodium potassium copper aluminium silicate hydroxide mineral. Ajoite has the chemical formula $(Na,K)Cu_7AlSi_9O_{24}(OH)_6 \cdot 3(H_2O)$, and minor Mn, Fe and Ca are usually also present in the structure. Ajoite is used as a minor ore of copper.

Discovery

In August 1941 Harry Berman of Harvard University was collecting at Ajo, in Pima County, Arizona, USA. He found specimens of dark blue shattuckite, together with a bluish green mineral which he suspected was a new species. Berman and W T Schaller had planned to collaborate on the investigation of this mineral, together with other known copper silicate minerals, but Berman died in a plane crash in 1944, aged 42, before this study was done. It was not until 1958 that Schaller, together with Angelina Vlisidis (both of the US Geological Survey) studied the greenish mineral and determined that it was indeed a new species. They named it "ajoite" (pronounced ah-hoe-ite) after the place where it had been found.

Twenty three years later, in 1981, George Chao from Carleton University in Ottawa, Canada, studied the mineral again, and showed that ajoite was triclinic, and not monoclinic as had been thought previously. His studies also resulted in a re-definition of the chemical formula.

Crystal Structure

The 1981 investigations of ajoite were done on a sample from the New Cornelia Mine, Ajo, with tiny crystals averaging 0.01×0.1×0.4 mm. The mineral was shown to belong to the triclinic system, but it was not clear whether the crystal class was 1 or 1. The triclinic class 1 has the lowest possible symmetry with no symmetry elements at all, and 1 has just a centre of symmetry, but no axes of symmetry or mirror planes. The space group is either P1 or P1.

The unit cell parameters are: a = 13.637 Å, b = 14.507 Å, c = 13.620 Å, α = 107.16°, β = 105.45° and ã = 110.57°.

Ajoite occurs as sprays of bladed prismatic crystals, commonly fibrous, elongated along the c crystal axis and flattened sideways. The dominant form is {010}, parallel to the plane containing the a and c axes. The forms {110}, cutting both the a and b axes, and {100}, parallel to the plane containing the b and c axes, are much less prominent but are always present. The termination on c may be either {001} or {203} or both.

Ajoite is frequently found as inclusions in quartz, much sought after by mineral collectors.

Optical Properties

Ajoite is translucent and bluish green in colour, with a pale greenish white streak and a vitreous luster. The optical class is biaxial (+) with refractive indices $N_x = 1.550$, $N_y = 1.583$ and $N_z = 1.641$. These values are similar to the refractive index of ordinary window glass, at 1.5. In white light the mineral is distinctly pleochroic, being very light bluish green along the X direction and brilliant bluish green along the Y and Z directions.

Physical properties

The cleavage is perfect parallel to the plane containing the a and c axes, and the hardness is 3.5, between that of calcite and fluorite. It is a little denser than quartz, with specific gravity 2.96. Ajoite is readily decomposed by hydrochloric acid, HCl, and by nitric acid, HNO_3, but it does not react with ammonium hydroxide, NH_4OH. It has been reported as having barely detectable radioactivity.

Environment

Ajoite is a secondary mineral that forms from the oxidation of other secondary copper minerals in copper-rich base metal deposits in massive fracture coatings, in vein fillings, and in vugs. It may form from shattuckite and also it may be replaced by shattuckite.

At the type locality it is associated with shattuckite, conichalcite, quartz, muscovite and pyrite.

Occurrence

Figure 16: Ajoite in quartz from South Africa (size: 1.7×1.4×1.2 cm)

Ajoite is named after its type locality, the New Cornelia Mine in the Ajo District of Pima County, Arizona.

Type material is conserved at the National Museum of Natural History, Washington DC, USA, reference number 113220.

Other localities include Wickenburg and Maricopa County in Arizona, within the United States, and the Messina District in South Africa.

Quartz specimens from the defunct Messina Mines on the border between Zimbabwe and South Africa are well known for their inclusions of blue copper silicate minerals such as shattuckite, papagoite and ajoite, but ajoite from American localities does not occur like this.

Akaganéite

Figure 17: A piece of the mineral akaganéite. Exhibit of the "Earth and Man" Museum in Sofia, Bulgaria. Discovered in Kaskasnyunchorr, Khibiny Massif, Kola Peninsula, Russia

Akaganéite is an iron(III) oxide-hydroxide / chloride mineral with formula: $Fe^{3+}O(OH,Cl)$ e.g; β-FeO(OH). It is formed by the weathering of pyrrhotite ($Fe_{1-x}S$). Nickel also has been reported in the structure. It has a monoclinic crystal structure, a metallic luster and a brownish yellow streak.

It is named after the Akagane mine in Iwate, Japan, in which it was discovered. It has also been found in widely dispersed locations around the world and in rocks from the Moon that were brought back during the Apollo Project. The occurrences in meteorites and the lunar sample are thought to have been produced by interaction with Earth's atmosphere.

Akdalaite

Akdalaite is a very rare mineral found in Kazakhstan and has the formula $5Al_2O_3 \cdot H_2O$. It was formerly believed to be $4Al_2O_3 \cdot H_2O$. It is therefore the same as tohdite an artificailly produced phase. Studies on the crystal structure and spectra indicate that this is an aluminium oxide hydroxide.

Akermanite

Akermanite ($Ca_2Mg[Si_2O_7]$) is a melilite mineral of the sorosilicate group, containing calcium, magnesium, silicon, and oxygen. It is a product of contact metamorphism of siliceous limestones and dolostones, and rocks of sanidinite facies.

Sanidinite facies represent the highest conditions of temperature of contact metamorphism and are characterised by the absence of hydrous minerals. It has a density of 2.944 g/cm^3.

Åkermanite ranks a 5 or 6 on the Mohs scale of mineral hardness, and can be found grey, green, brown, or colourless. It has a white streak and a vitreous or resinous luster.

It has a tetragonal crystal system and a good, or distinct, cleavage. It is the end member in a solid solution series beginning with gehlenite ($Ca_2Al[AlSiO_2]$).

The mineral is named for Anders Richard Åkerman, a Swedish metallurgist. It has been found at Monte Somma and Vesuvius, and Monte Cavalluccio near Rome. It was "grandfathered" in as a species of mineral because it was described prior to 1959, before the founding of the International Mineralogical Association.

Alabandite

Figure 18: Alabandite from Peru

Alabandite or alabandine is a rarely occurring manganese sulfide mineral. It crystallises in the cubic crystal system with the chemical composition $Mn^{2+}S$ and develops commonly massive to granular aggregates, but rarely also cubic or octahedral crystals to 1 cm.

Etymology and History

Alabandite was first described in 1784 by Franz-Joseph Müller von Reichenstein. The mineral name is derived from its supposed discovery locality at Alabanda (Aïdin) in Turkey.

Occurrence

Alabandite forms in epithermal polymetallic sulfide veins and low-temperature manganese deposits. It occurs with acanthite, calcite, chalcopyrite, galena, pyrite, quartz, rhodochrosite, rhodonite, sphalerite and native tellurium. Sometimes it was found in meteorites. Localities are several areas in Antarctica, Argentina, Armenia, Australia, Bolivia, Brazil, Bulgaria, Canada, China, Czech Republic, Finland, France, Germany, Ghana, Greece, Greenland, India, Italy, Japan, Kyrgyzstan, Mexico, New Zealand, Norway, Peru, Poland, Romania, Russia, Slovakia, South Africa, South Korea, Sweden, Switzerland, Taiwan, Tanzania, the United Kingdom, the USA, Uzbekistan and Yemen. Altogether at present time approximately 220 discovery sites are registered.

Crystal Structure

Alabandite crystallises in the cubic crystal system in the space group Fm3m with the lattice parameter a = 5.22 Å and four formula units per unit cell.

Alamosite

Figure 19: Transparent crystals of alamosite in the matrix of leadhillite and melanotekite

Alamosite ($Pb_{12}Si_{12}O_{36}$) is a colourless silicate mineral named after the place where it was discovered, Álamos, Sonora, Mexico. It is a rare secondary mineral occurring in the oxidised zones of lead-rich deposits. For example, the infobox picture shows its association with black leadhillite.

Albite

Figure 20: Albite from Crete, scale = 1 in.

Albite is a plagioclase feldspar mineral. It is the sodium endmember of the plagioclase solid solution series. As such it represents a plagioclase with less than 10% anorthite content. The pure albite endmember has the formula $NaAlSi_3O_8$. It is a tectosilicate. Its colour is usually pure white, hence its name from Latin *albus*.

Figure 21: Albite

Albite crystallises with triclinic pinacoidal forms. Its specific gravity is about 2.62 and it has a Mohs hardness of 6 - 6.5. Albite almost always exhibits crystal twinning often as minute parallel striations on the crystal face. Albite often occurs as fine parallel segregations alternating with pink microcline in perthite as a result of exolution on cooling.

It occurs in granitic and pegmatite masses, in some hydrothermal vein deposits and forms part of the typical greenschist metamorphic facies for rocks of originally basaltic composition.

It was first reported in 1815 for an occurrence in Finnbo, Falun, Dalarna, Sweden. The name is from Latin, *albus* for the typical white colour.

Alforsite

Alforsite is a mineral, $Ba_5(PO_4)_3Cl$, composed of barium, phosphorus, chlorine, and oxygen. It was discovered in 1981, and named to honour geologist John T. Alfors (1930-2005) of the California Geological Survey (formerly California Division of Mines and Geology) for his work in the area where it was discovered.

Alforsite is a hexagonal colourless crystal in the chemical class phosphates and the group apatite. It is found in certain parts of central California, primarily Fresno, Mariposa, and Tulare Counties. It has also been found in Baja California, Mexico.

Algodonite

Figure 22: Mohawkite nugget, a mixture of domeykite, algodonite and copper

Algodonite is a copper arsenide mineral with formula: Cu_6As. It is a gray white metallic mineral crystallising in the hexagonal system. It has a Mohs hardness of 4 and a specific gravity of 8.38 - 8.72.

It was first described in 1857 from the Algodones silver mine, Coquimbo, Chile.

Aliettite

Figure 23: Aliettite

Aliettite is a complex phyllosilicate mineral of the smectite group with a formula of $(Ca_{0.2}Mg_6(Si,Al)_8O_{20}(OH)_4 \cdot 4H_2O)$ or $[Mg_3Si_4O_{10}(OH)_2](Ca_{0.5},Na)_{0.33}(Al,Mg,Fe^{2+})_{2-3}(Si,Al)_4O_{10}(OH)_2 \cdot n(H_2O)$.

It is a soft, colourless to pale yellow or green earthy mineral which crystallises in the monoclinic system as minute tabular to platy crystals.

It was first described in 1968 for an occurrence in Monte Chiaro, Albareto, Parma Province, Emilia-Romagna, Italy and named for the Italian mineralogist Andrea Alietti.

It occurs in serpentinized ophiolites and their residual soil. It also occurs in altered dolostone. Associated minerals include talc, chlorite, serpentine and calcite. In addition tot the type locality in Italy it has been reported from Kinshasa, Katanga; the Chelyabinsk Oblast of the southern Urals and the Turii alkaline Massif of the Kola Peninsula in Russia; the Zirabulak Mountains of Uzbekistan; and the Goldstrike Mine of Eureka County, Nevada, USA.

Allabogdanite

Allabogdanite is a very rare phosphide mineral with formula $(Fe,Ni)_2P$, found in 1997 in a meteorite. It was described for an occurrence in the Onello meteorite in the Onello River basin, Sakha Republic; Yakutia, Russia; associated with taenite, schreibersite, kamacite, graphite and awaruite. It was named for Russian geologist Alla Bogdanova.

Allanite

Figure 24: Allanite from the Mt. Isa – Cloncurry area, Queensland, Australia (scale bar 1 inch)

Allanite is a sorosilicate group of minerals within the broader epidote group that contain a significant amount of rare earth elements. The mineral occurs mainly in metamorphosed clay rich sediments and felsic igneous rocks. It has the general formula $A_2M_3Si_3O_{12}[OH]$ where the A sites can contain large cations such as Ca^{2+}, Sr^{2+}, and rare earth elements and the M sites admit Al^{3+}, Fe^{3+}, Mn^{3+}, Fe^{2+}, or Mg^{2+} among others. However, a large amount of additional elements, including Th, U, Zr, P, Ba, Cr and others may be present in the mineral. The International Mineralogical Association lists three minerals in the allanite group, each recognised as a unique mineral: allanite-(Ce), allanite-(La) and allanite-(Y), depending on the dominant rare earth present, cerium, lanthanum or yttrium.

Allanite, also called orthite, contains up to 20% rare earth elements and is a valuable source of them. The inclusion of thorium and other radioactive elements in allanite results in some interesting phenomena. Allanite often has a pleochroic halo of radiation damage in the minerals immediately adjacent. Also highly radioactive grains of allanite often have their structure disrupted or are *metamict.* The age of allanite grains that have not been destroyed by radiation can be determined using different techniques.

Figure 25: Allanite crystals on smokey quartz from the White Mountain Wilderness, Lincoln County, New Mexico, USA (size: 2.7×1.8×1.7 cm)

Allanite is usually black in colour, but can be brown, brown-violet. It is often coated with a yellow-brown alertation product, likely limonite. It crystallises in the monoclinic system and forms prismatic crystals. It has a Mohs hardness of 5.5–6 and a specific gravity of 3.5–4.2.

It is also pyrognomic, meaning that it becomes incandescent at a relatively low temperature of about 95 °C.

It was discovered in 1810 and named for the Scottish mineralogist, Thomas Allan (1777–1833). The type locality is Aluk Island, Greenland where it was first discovered by Karl Ludwig Giesecke.

Alloclasite

Alloclasite ((Co,Fe)AsS) is a sulfosalt mineral. It is a member of the arsenopyrite group. Alloclasite crystallises in the monoclinic system and typically forms as columnar to radiating acicular prismatic clusters. It is an opaque steel-gray to silver-white, with a metallic lustre and a black streak. It is brittle with perfect cleavage, a Mohs hardness of 5 and a specific gravity of 5.91–5.95.

It was first described in 1866 for an occurrence in Romania. Its name is derived from Greek for "other" and "to break," in reference to its distinct cleavage which distinguished it from the similar appearing mineral marcasite.

Allophane

Allophane is a poorly-crystalline hydrous aluminium silicate clay *mineraloid.* Its chemical formula is $Al_2O_3 \cdot (SiO_2)_{1.3\text{-}2} \cdot 2.5\text{-}3(H_2O)$. Since it has short-range atomic order, it is a mineraloid, rather than a mineral, and can be identified by its distinctive infrared spectrum and its X-ray diffraction pattern. It was first described in 1816 in Gräfenthal, Thuringia, Germany.

Allophane is a weathering or hydrothermal alteration product of volcanic glass and feldspars and has a composition similar to kaolinite. It typically forms under mildly acidic to neutral pH (5-7). Its structure has been debated, but it is similar to clay minerals and is composed of curved alumina octahedral and silica tetrahedral layers. Transmission electron micrographs show that it is generally made up of aggregates of hollow spherules ~3-5 nm in diametre. Allophane can alter to form halloysite under resilicating aqueous conditions and can alter to form gibbsite under desilicating conditions. A copper containing variety *cupro-allophane* has been reported.

It forms waxy botryoidal to crusty masses with colour varying from white through green, blue, yellow, to brown. It has a Mohs hardness of 3 and a specific gravity of 1.0.

It was named from the Greek *allos* - "other" and *phanos* - "to appear", as it gave a deceptive reaction in the blowpipe flame in old mineralogical testing.

Almandine

Almandine also known incorrectly as almandite, is a species of mineral belonging to the garnet Group. The name is a corruption of alabandicus, which is the name applied by Pliny the Elder to a stone found or worked at Alabanda, a town in Caria in Asia Minor. Almandine is an iron alumina garnet, of deep red colour, inclining to purple. It is frequently cut with a convex face, or en cabochon, and is then known as carbuncle. Viewed through the spectroscope in a strong light, it generally shows three characteristic absorption bands.

Figure 26: Almandine

Almandine is one end-member of a mineral solid solution series, with the other end member being the garnet pyrope. The almandine crystal formula is: $Fe_3Al_2(SiO_4)_3$. Magnesium substitutes for the iron with increasingly pyrope-rich composition.

Almandine, $Fe^{2+}{}_3Al_2Si_3O_{12}$, is the ferrous iron end member of the class of garnet minerals representing an important group of rock-forming silicates, which are the main constituents of the Earth's crust, upper mantle and transition zone. Almandine crystallises in the cubic space group Ia3d, with unit-cell parameter a H⁻ 11.512 Å at 100 K.

Almandine is antiferromagnet with the Néel temperature of 7.5 K. It contains two equivalent magnetic sublattices

Occurrence

Figure 27: Almandine

Almandine occurs rather abundantly in the gem-gravels of Sri Lanka, whence it has sometimes been called Ceylon-ruby. When the colour inclines to a violet tint, the stone is often called Syrian garnet, a name said to be taken from Syriam, an ancient town of Pegu. Large deposits of fine almandine-garnets were found, some years ago, in the Northern Territory of Australia, and were at first taken for rubies and thus they were known in trade for some time afterwards as Australian rubies.

Almandine is widely distributed. Fine rhombic dodecahedra occur in the schistose rocks of the Zillertal, in Tyrol, and are sometimes cut and polished. An almandine in which the ferrous oxide is replaced partly by magnesia is found at Luisenfeld in German East Africa. In the United States there are many localities which yield almandine. Fine crystals of almandine embedded in mica-schist occur near Fort Wrangell in Alaska. The coarse varieties of almandine are often crushed for use as an abrasive agent.

Alstonite

Figure 28: Alstonite crystals from Brownley Hill Mine, North Pennines, Cumbria, England (size: 5.1 x 4.1 x 2.7 cm)

Alstonite, also known as bromlite, is a member of the aragonite group of carbonate minerals. It is a calcium and barium carbonate mineral with formula, $BaCa(CO_3)_2$. It is triclinic in crystal structure and differs structurally from barytocalcite, which has a monoclinic structure. Simple crystals are not known. The crystals are invariably complex twins, and have the form of doubly terminated pseudohexagonal pyramids, like those of witherite but more acute; the faces are horizontally striated and are divided down their centre by a twin suture.

The examination in polarised light of a transverse section shows that each compound crystal is built up of six differently oriented individuals arranged in twelve segments.

The crystals are translucent and white, sometimes with a shade of pink. Specific gravity is 3.706, and hardness is 4 - 4.1.

The mineral has been found in a variety of places, two of which are in the north of England. At the Fallowfield lead mine, near Hexham in Northumberland, it is associated with witherite; and at Bromley Hill, near Alston in Cumberland, it occurs in veins with galena. The species was named bromlite by T. Thomson in 1837, and alstonite by A. Breithaupt in 1841, both of which names, derived from the locality, have been in common use.

It occurs typically in low-temperature hydrothermal lead-zinc ore deposits and has been reported as a rare phase in carbonatites. It occurs associated with calcite, baryte, ankerite, siderite, benstonite, galena, sphalerite, pyrite and quartz.

Altaite

Altaite, or lead telluride, is a yellowish white mineral with an isometric crystal structure. Altaite is in the galena group of minerals as it shares many of properties of galena. Altaite has an unusually high density for a light-coloured mineral. Altaite and other rare tellurides are classified in the sulfide mineral class (Dana classification).

Altaite was discovered in 1845 in the Altai Mountains. Besides these mountains altaite can also be found in Zyrianovsk, Kazakhstan; the Ritchie Creek Deposit in Price County, Wisconsin; the Koch-Bulak gold deposit in Kazakhstan; Moctezuma, Mexico; and Coquimbo, Chile among other locations.

Aluminite

Aluminite is a hydrous aluminium sulfate mineral with formula: $Al_2SO_4(OH)_4 \cdot 7(H_2O)$. It is an earthy white to gray-white monoclinic mineral which almost never exhibits crystal form. It forms botryoidal to mammillary clay-like masses. It is very soft Mohs hardness of 1 with a specific gravity of 1.66—1.7. It forms as an oxidation product of pyrite and marcasite along with aluminous silicates and clays. It was first described in 1807 from Halle, Saxony-Anhalt, Germany. It is also known as *alley stone* and *websterite* (named after English geologist Thomas Webster).

Aluminite is used by tile & masonry workers to reduce the setting time of cementitious mortars.

Aluminium

Figure 29: Spectral lines of aluminium

Aluminium or aluminium (US English) is a silvery white member of the boron group of chemical elements. It has the symbol Al, and its atomic number is 13. It is not soluble in water under normal circumstances. Aluminium is the third most abundant element (after oxygen and silicon), and the most abundant metal, in the Earth's crust. It makes up about 8% by weight of the Earth's solid surface. Aluminium metal is too reactive chemically to occur natively. Instead, it is found combined in over 270 different minerals. The chief ore of aluminium is bauxite.

Aluminium is remarkable for the metal's low density and for its ability to resist corrosion due to the phenomenon of passivation. Structural components made from aluminium and its alloys are vital to the aerospace industry and are important in other areas of transportation and structural materials. The most useful compounds of aluminium, at least on a weight basis, are the oxides and sulfates.

Despite its prevalence in the environment, aluminium salts are not known to be used by any form of life. In keeping with its pervasiveness, it is well tolerated by plants and animals. Because of their prevalence, potential biological roles, beneficial and otherwise, aluminium compounds are of continuing interest.

Characteristics

Figure 30: Etched surface from a high purity (99.9998%) aluminium bar, size 55×37 mm

Physical

Aluminium is a soft, durable, lightweight, ductile and malleable metal with appearance ranging from silvery to dull gray, depending on the surface roughness. Aluminium is nonmagnetic and does not easily ignite. A fresh film of aluminium film serves as a good reflector (approximately 92%) of visible light and an excellent reflector (as much as 98%) of medium and far infrared radiation. The yield strength of pure aluminium is 7–11 MPa, while aluminium alloys have yield strengths ranging from 200 MPa to 600 MPa. Aluminium has about one-third the density and stiffness of steel. It is easily machined, cast, drawn and extruded.

Corrosion resistance can be excellent due to a thin surface layer of aluminium oxide that forms when the metal is exposed to air, effectively preventing further oxidation. The strongest aluminium alloys are less corrosion resistant due to galvanic reactions with alloyed copper. This corrosion resistance is also often greatly reduced when many aqueous salts are present, particularly in the presence of dissimilar metals.

Aluminium atoms are arranged in a face-centred cubic (fcc) structure. Aluminium has a stacking-fault energy of approximately 200 mJ/m^2.

Aluminium is one of the few metals that retain full silvery reflectance in finely powdered form, making it an important component of silver-coloured paints. Aluminium mirror finish has the highest reflectance of any metal in the 200–400 nm (UV) and the 3,000–10,000 nm (far IR) regions; in the 400–700 nm visible range it is slightly outperformed by tin and silver and in the 700–3000 (near IR) by silver, gold, and copper.

Aluminium is a good thermal and electrical conductor, having 59% the conductivity of copper, both thermal and electrical. Aluminium is capable of being a superconductor, with a superconducting critical temperature of 1.2 Kelvin and a critical magnetic field of about 100 gauss (10 milliteslas).

Creation

Stable aluminium is created when hydrogen fuses with magnesium either in large stars or in supernovae.

Isotopes

Aluminium has many known isotopes, whose mass numbers range from 21 to 42; however, only ^{27}Al (stable isotope) and ^{26}Al (radioactive isotope, $t_{1/2} = 7.2 \times 10^5$ y) occur naturally. ^{27}Al has a natural abundance above 99.9%. ^{26}Al is produced from argon in the atmosphere by spallation caused by cosmic-ray protons. Aluminium isotopes have found practical application in dating marine sediments, manganese nodules, glacial ice, quartz in rock exposures, and meteorites. The ratio of ^{26}Al to ^{10}Be has been used to study the role of transport, deposition, sediment storage, burial times, and erosion on 10^5 to 10^6 year time scales. Cosmogenic ^{26}Al was first applied in studies of the Moon and meteorites. Meteoroid fragments, after departure from their parent bodies, are exposed to intense cosmic-ray bombardment during their travel through space, causing substantial ^{26}Al production. After falling to Earth, atmospheric shielding drastically reduces ^{26}Al production, and its decay can then be used to determine the meteorite's terrestrial age. Meteorite research has also shown that ^{26}Al was relatively abundant at the time of formation of our planetary system. Most meteorite scientists believe that the energy released by the decay of ^{26}Al was responsible for the melting and differentiation of some asteroids after their formation 4.55 billion years ago.

Natural Occurrence

In the Earth's crust, aluminium is the most abundant (8.3% by weight) metallic element and the third most abundant of all elements (after oxygen and silicon). Because of its strong affinity to oxygen, it is almost never found in the elemental state; instead it is found in oxides or silicates. Feldspars, the most common group of minerals in the Earth's crust, are aluminosilicates. Native aluminium metal can be found as a minor phase in low oxygen fugacity environments, such as the interiors of certain volcanoes. Native aluminium has been reported in cold seeps in the northeastern continental slope of the South China Sea and Chen et al. (2011) have proposed a theory of its origin as resulting by reduction from tetrahydroxoaluminate $Al(OH)_4^-$ to metallic aluminium by bacteria.

It also occurs in the minerals beryl, cryolite, garnet, spinel and turquoise. Impurities in Al_2O_3, such as chromium or iron yield the gemstones ruby and sapphire, respectively.

Although aluminium is an extremely common and widespread element, the common aluminium minerals are not economic sources of the metal. Almost all metallic aluminium is produced from the ore bauxite ($AlO_x(OH)_{3-2x}$). Bauxite occurs as a weathering product of low iron and silica bedrock in tropical climatic conditions. Large deposits of bauxite occur in Australia, Brazil, Guinea and Jamaica and the primary mining areas for the ore are in Australia, Brazil, China, India, Guinea, Indonesia, Jamaica, Russia and Surinam.

Production and Refinement

Although aluminium is the most abundant metallic element in the Earth's crust, it is never found in free, metallic form (with above mentioned exceptions).

Aluminium forms strong chemical bonds with oxygen. Compared to most other metals, it is difficult to extract from ore, such as bauxite, due to the energy required to reduce aluminium oxide (Al_2O_3). For example, direct reduction with carbon, as is used to produce iron, is not chemically possible, since aluminium is a stronger reducing agent than carbon. There is an indirect carbothermic reduction possible by using carbon and Al_2O_3, which forms an intermediate Al_4C_3 and this can further yield aluminium metal at a temperature of 1900–2000°C. This process is still under development.

Figure 31: Bauxite, a major aluminium ore. The red-brown colour is due to the presence of iron minerals.

This process costs less energy and yields less CO_2 than the Hall-Héroult process, the major industrial process for aluminium extraction. Aluminium oxide has a melting point of about 2,000 °C (3,600 °F). Therefore, it must be extracted by electrolysis.

In this process, the aluminium oxide is dissolved in molten cryolite with calcium fluoride and then reduced to the pure metal. The operational temperature of the reduction cells is around 950 to 980 °C (1,740 to 1,800 °F).

Cryolite is found as a mineral in Greenland, but in industrial use it has been replaced by a synthetic substance. Cryolite is a chemical compound of aluminium and sodium fluorides: (Na_3AlF_6). The aluminium oxide (a white powder) is obtained by refining bauxite in the Bayer process of Karl Bayer. (Previously, the Deville process was the predominant refining technology.)

The electrolytic process replaced the Wöhler process, which involved the reduction of anhydrous aluminium chloride with potassium. Both of the electrodes used in the electrolysis of aluminium oxide are carbon. Once the refined alumina is dissolved in the electrolyte, its ions are free to move around. The reaction at the cathode is:

$$Al^{3+} + 3\ e^- \rightarrow Al$$

Here the aluminium ion is being reduced. The aluminium metal then sinks to the bottom and is tapped off, usually cast into large blocks called aluminium billets for further processing.

At the anode, oxygen is formed:

$$2\ O^{2-} \rightarrow O_2 + 4\ e^-$$

To some extent, the carbon anode is degraded by the oxygen. The anodes in a reduction cell must therefore be replaced regularly, since they are consumed in the process. The cathodes do erode, mainly due to electrochemical processes and metal movement. After five to ten years, depending on the current used in the electrolysis, a cell has to be rebuilt because of cathode wear.

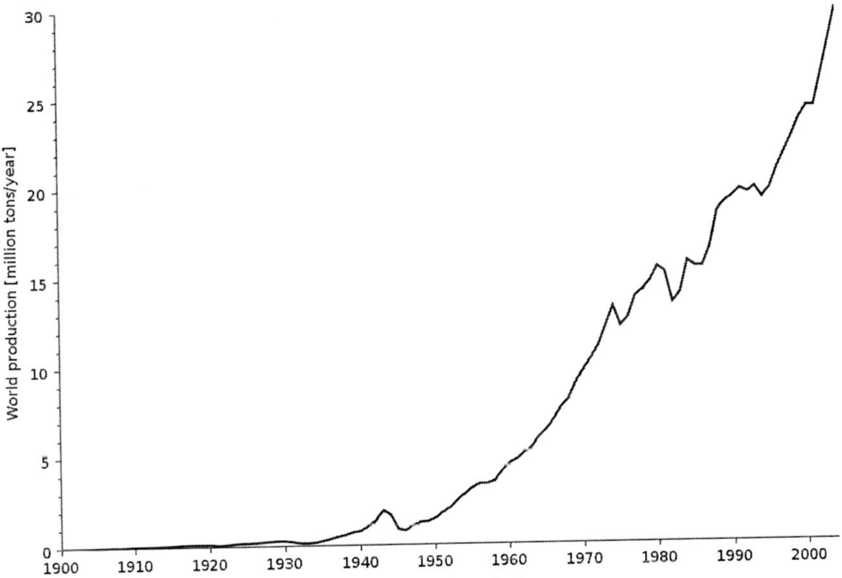

Figure 32: World production trend of aluminium

Aluminium electrolysis with the Hall-Héroult process consumes a lot of energy, but alternative processes were always found to be less viable economically and/or ecologically. The worldwide average specific energy consumption is approximately 15±0.5 kilowatt-hours per kilogram of aluminium produced (52 to 56 MJ/kg). The most modern smelters achieve approximately 12.8 kW·h/kg (46.1 MJ/kg). (Compare this to the heat of reaction, 31 MJ/kg, and the Gibbs free energy of reaction, 29 MJ/kg.) Reduction line currents for older technologies are typically 100 to 200 kiloamperes; state-of-the-art smelters operate at about 350 kA. Trials have been reported with 500 kA cells.

The Hall-Heroult process produces aluminium with a purity of above 99%. Further purification can be done by the Hoope process. The process involves the electrolysis of molten aluminium with a sodium, barium and aluminium fluoride electrolyte. The resulting aluminium has a purity of 99.99%.

Electric power represents about 20% to 40% of the cost of producing aluminium, depending on the location of the smelter. Aluminium production consumes roughly 5% of electricity generated in the U.S. Smelters tend to be situated where electric power is both plentiful and inexpensive, such as the United Arab Emirates with excess natural gas supplies and Iceland and Norway with energy generated from renewable sources. The world's largest smelters of alumina are People's Republic of China, Russia, and Quebec and British Columbia in Canada.

In 2005, the People's Republic of China was the top producer of aluminium with almost a one-fifth world share, followed by Russia, Canada, and the USA, reports the British Geological Survey.

Over the last 50 years, Australia has become a major producer of bauxite ore and a major producer and exporter of alumina (before being overtaken by China in 2007). Australia produced 62 million tonnes of bauxite in 2005. The Australian deposits have some refining problems, some being high in silica, but have the advantage of being shallow and relatively easy to mine.

Recycling

Figure 33: Aluminium recycling code

Aluminium is theoretically 100% recyclable without any loss of its natural qualities. According to the International Resource Panel's

Metal Stocks in Society report, the global per capita stock of aluminium in use in society (i.e. in cars, buildings, electronics etc.) is 80 kg. Much of this is in more-developed countries (350–500 kg per capita) rather than less-developed countries (35 kg per capita). Knowing the per capita stocks and their approximate lifespans is important for planning recycling.

Recovery of the metal via recycling has become an important facet of the aluminium industry. Recycling was a low-profile activity until the late 1960s, when the growing use of aluminium beverage cans brought it to the public awareness.

Recycling involves melting the scrap, a process that requires only 5% of the energy used to produce aluminium from ore, though a significant part (up to 15% of the input material) is lost as dross (ash-like oxide). The dross can undergo a further process to extract aluminium.

In Europe aluminium experiences high rates of recycling, ranging from 42% of beverage cans, 85% of construction materials and 95% of transport vehicles.

Recycled aluminium is known as secondary aluminium, but maintains the same physical properties as primary aluminium. Secondary aluminium is produced in a wide range of formats and is employed in 80% of alloy injections. Another important use is for extrusion.

White dross from primary aluminium production and from secondary recycling operations still contains useful quantities of aluminium that can be extracted industrially. The process produces aluminium billets, together with a highly complex waste material. This waste is difficult to manage. It reacts with water, releasing a mixture of gases (including, among others, hydrogen, acetylene, and ammonia), which spontaneously ignites on contact with air; contact with damp air results in the release of copious quantities of ammonia gas. Despite these difficulties, the waste has found use as a filler in asphalt and concrete.

Compounds

Oxidation State +3: The vast majority of compounds, including all Al-containing minerals and all commercially significant aluminium compounds, feature aluminium in the oxidation state 3+. The coordination number of such compounds varies, but generally Al^{3+} is six-coordinate or tetracoordinate. Almost all compounds of aluminium(III) are colourless.

Halides

All four trihalides are well known. Unlike the structures of the three heavier trihalides, aluminium fluoride (AlF_3) features six-coordinate Al. The octahedral coordination environment for AlF_3 is related to the compactness of fluoride ion, six of which can fit around the small Al^{3+} centre. AlF_3 sublimes (with cracking) at 1,291 °C (2,356 °F). With heavier halides, the coordination numbers are lower. The other trihalides are dimeric or polymeric with tetrahedral Al centres. These materials are prepared by treating aluminium metal with the halogen, although other methods exist. Acidification of the oxides or hydroxides affords hydrates. In aqueous solution, the halides often form mixtures, generally containing six-coordinate Al centres, which are feature both halide and aquo ligands. When aluminium and fluoride are together in aqueous solution, they readily form complex ions such as $[AlF(H_2O)_5]^{2+}$, $AlF_3(H_2O)_3$, and $[AlF_6]^{3-}$. In the case of chloride, polyaluminium clusters are formed such as $[Al_{13}O_4(OH)_{24}(H_2O)_{12}]^{7+}$.

Oxide and Hydroxides

Aluminium forms one stable oxide, known by its mineral name corundum. Sapphire and ruby are impure corundum contaminated with trace amounts of other metals. The two oxide-hydroxides (AlO(OH) are boehmite and diaspore. There are three trihydroxides: bayerite, gibbsite, and nordstrandite, which differ in their crystalline structure (polymorphs). Most are produced from ores by a variety of wet processes using acid and base. Heating the hydroxides leads to formation of corundrum. These materials are of central importance to the production of aluminium and are themselves extremely useful.

Carbide, Nitride, and Related Materials

Aluminium carbide (Al_4C_3) is made by heating a mixture of the elements above 1,000 °C (1,832 °F). The pale yellow crystals consist of tetrahedral aluminium centres. It reacts with water or dilute acids to give methane. The acetylide, $Al_2(C_2)_3$, is made by passing acetylene over heated aluminium.

Aluminium nitride (AlN) is the only nitride known for aluminium. Unlike the oxides it features tetrahedral Al centres. It can be made from the elements at 800 °C (1,472 °F). It is air-stable material with a usefully high thermal conductivity. Aluminium phosphide (AlP) is made similarly, and hydrolyses to give phosphine:

$$AlP + 3\ H_2O \rightarrow Al(OH)_3 + PH_3$$

Organoaluminium Compounds and Related Hydrides

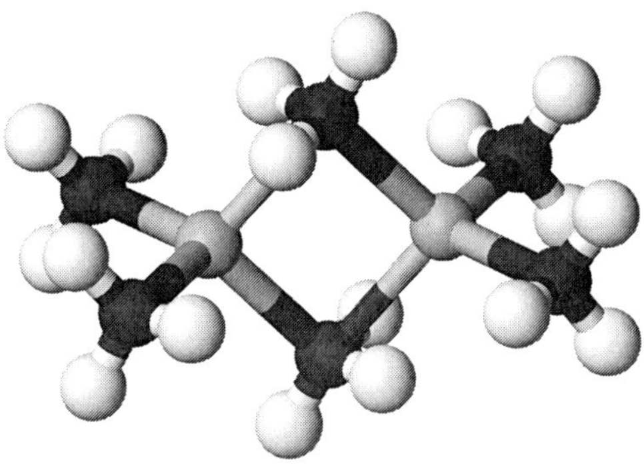

Figure 34: Structure of trimethylaluminium, a compound that features five-coordinate carbon.

A variety of compounds of empirical formula AlR_3 and $AlR_{1.5}Cl_{1.5}$ exist. These species usually feature tetrahedral Al centres, e.g. "trimethylaluminium" has the formula $Al_2(CH_3)_6$.

With large organic groups, triorganoaluminium exist as three-coordinate monomers, such as triisobutylaluminium.

Such compounds are widely used in industrial chemistry, despite the fact that they are often highly pyrophoric. Few analogues exist between organoaluminium and organoboron compounds except for large organic groups.

The important aluminium hydride is lithium aluminium hydride ($LiAlH_4$), which is used in as a reducing agent in organic chemistry. It can be produced from lithium hydride and aluminium trichloride:

$$4\ LiH + AlCl_3 \rightarrow LiAlH_4 + 3\ LiCl$$

Several useful derivatives of $LiAlH_4$ are known, e.g. Sodium bis(2-methoxyethoxy)dihydridoaluminate. The simplest hydride, aluminium hydride or alane, remains a laboratory curiosity. It is a polymer with the formula $(AlH_3)_n$, in contrast to the corresponding boron hydride with the formula $(BH_3)_2$.

Oxidation States +1 and +2

Although the great majority of aluminium compounds feature Al^{3+} centres, compounds with lower oxidation states are known and sometime of significance as precursors to the Al^{3+} species.

Aluminium(I)

AlF, AlCl and AlBr exist in the gaseous phase when the tri-halide is heated with aluminium. The composition AlI is unstable at room temperature with respect to the triiodide:

$$3\ AlI \rightarrow AlI_3 + 2\ Al$$

A stable derivative of aluminium monoiodide is the cyclic adduct formed with triethylamine, $Al_4I_4(NEt_3)_4$. Also of theoretical interest but only of fleeting existence are Al_2O and Al_2S is made by heating the normal oxide, Al_2O_3, with silicon at 1,800 °C (3,272 °F) in a vacuum. Such materials quickly disproportionates to the starting materials.

Aluminium(II)

Very simple Al(II) compounds are invoked or observed in the reactions of Al metal with oxidants. For example, aluminium monoxide, AlO, has been detected in the gas phase after explosion and in stellar absorption spectra. More thoroughly investigated are compounds of the formula R_4Al_2 where R is a large organic ligand.

Analysis

The presence of aluminium can be detected in qualitative analysis using aluminon.

Applications

General Use: Aluminium is the most widely used non-ferrous metal. Global production of aluminium in 2005 was 31.9 million tonnes. It exceeded that of any other metal except iron (837.5 million tonnes). Forecast for 2012 is 42–45 million tons, driven by rising Chinese output.

Aluminium is almost always alloyed, which markedly improves its mechanical properties, especially when tempered. For example, the common aluminium foils and beverage cans are alloys of 92% to 99% aluminium. The main alloying agents are copper, zinc, magnesium, manganese, and silicon (e.g., duralumin) and the levels of these other metals are in the range of a few percent by weight.

Some of the many uses for aluminium metal are in:

- Transportation (automobiles, aircraft, trucks, railway cars, marine vessels, bicycles, etc.) as sheet, tube, castings, etc.
- Packaging (cans, foil, etc.)
- Construction (windows, doors, siding, building wire, etc.)
- A wide range of household items, from cooking utensils to baseball bats, watches.
- Street lighting poles, sailing ship masts, walking poles, etc.
- Outer shells of consumer electronics, also cases for equipment e.g. photographic equipment.
- Electrical transmission lines for power distribution
- MKM steel and Alnico magnets
- Super purity aluminium (SPA, 99.980% to 99.999% Al), used in electronics and CDs.
- Heat sinks for electronic appliances such as transistors and CPUs.
- Substrate material of metal-core copper clad laminates used in high brightness LED lighting.
- Powdered aluminium is used in paint, and in pyrotechnics such as solid rocket fuels and thermite.
- Aluminium can be reacted with hydrochloric acid or with sodium hydroxide to produce hydrogen gas.
- A variety of countries, including France, Italy, Poland, Finland, Romania, Israel, and the former Yugoslavia, have issued coins struck in aluminium or aluminium-copper alloys.
- Some guitar models sports aluminium diamond plates on the surface of the instruments, usually either chrome or black. Kramer Guitars and Travis Bean are both known for having produced guitars with necks made of aluminium, which gives the instrument a very distinct sound.
- Sustainability of Aluminium in Buildings.

Relatively pure aluminium is rarely encountered – only when corrosion resistance and/or workability is more important than strength or hardness. A thin layer of aluminium can be deposited onto a flat surface by physical vapour deposition or (very infrequently) chemical vapour deposition or other chemical means to form optical coatings and mirrors.

Aluminium Compounds

Because aluminium is abundant and most of its derivative exhibit low toxicity, the compounds of aluminium enjoy wide and sometimes large-scale applications.

Alumina

Aluminium oxide (Al_2O_3) and the associated oxy-hydroxides and trihydroxides are produced or extracted from minerals on a large scale. The great majority of this material is converted to metallic aluminium. About 10% of the production capacity is used for other applications. A major use is as an absorbent, for example alumina will remove water from hydrocarbons, to enable subsequent processes that are poisoned by moisture. Aluminium oxides are common catalysts for industrial processes, e.g. the Claus process for converting hydrogen sulfide to sulfur in refineries and for the alkylation of amines. Many industrial catalysts are "supported", meaning generally that an expensive catalyst (e.g., platinum) is dispersed over a high surface area material such as alumina. Being a very hard material (Mohs hardness 9), alumina is widely used as an abrasive and the production of applications that exploit its inertness, e.g., in high pressure sodium lamps.

Sulfates

Several sulfates of aluminium find applications. Aluminium sulfate ($Al_2(SO_4)_3(H_2O)_{18}$) is produced on the annual scale of several billions of kilograms. About half of the production is consumed in water treatment. The next major application is in the manufacture of paper. It is also used as a mordant, in fire extinguisher, as a food additive, in fireproofing, and in leather tanning. Aluminium ammonium sulfate ($[Al(NH_4)](SO_4)_2)(H_2O)_{12}$, also called ammonium alum, is used as a mordant and in leather tanning. Aluminium potassium sulfate ($[Al(K)](SO_4)_2)(H_2O)_{12}$ is used similarly. The consumption of both alums is declining.

Chlorides

Aluminium chloride ($AlCl_3$) is used in petroleum refining and in the production of synthetic rubber and polymers. Although it has a similar name, aluminium chlorohydrate has fewer and very different applications, e.g. as an hardening agent and an antiperspirant. It is an intermediate in the production of aluminium metal.

Niche Compounds

Given the scale of aluminium compounds, a small scale application could still involve thousands of tons. One of the many compounds used at this intermediate level include aluminium acetate, a salt used in solution as an astringent.

Aluminium borate (Al_2O_3 B_2O_3) is used in the production of glass and ceramics. Aluminium fluorosilicate ($Al_2(SiF_6)_3$) is used in the production of synthetic gemstones, glass and ceramic. Aluminium phosphate ($AlPO_4$) is used in the manufacture: of glass and ceramic, pulp and paper products, cosmetics, paints and varnishes and in making dental cement.

Aluminium hydroxide ($Al(OH)_3$) is used as an antacid, as a mordant, in water purification, in the manufacture of glass and ceramic and in the waterproofing of fabrics. Lithium aluminium hydride is a powerful reducing agent used in organic chemistry. Organoaluminiums are used as Lewis acids and cocatalysts. For example, methylaluminoxane is a cocatalyst for Ziegler-Natta olefin polymerisation to produce vinyl polymers such as polyethene.

Aluminium Alloys in Structural Applications

Figure 35: Aluminium foam

Aluminium alloys with a wide range of properties are used in engineering structures. Alloy systems are classified by a number system (ANSI) or by names indicating their main alloying constituents (DIN and ISO).

The strength and durability of aluminium alloys vary widely, not only as a result of the components of the specific alloy, but also as a result of heat treatments and manufacturing processes. A lack of knowledge of these aspects has from time to time led to improperly designed structures and gained aluminium a bad reputation.

One important structural limitation of aluminium alloys is their fatigue strength. Unlike steels, aluminium alloys have no well-defined fatigue limit, meaning that fatigue failure eventually occurs, under even very small cyclic loadings. This implies that engineers must assess these loads and design for a fixed life rather than an infinite life.

Another important property of aluminium alloys is their sensitivity to heat. Workshop procedures involving heating are complicated by the fact that aluminium, unlike steel, melts without first glowing red. Forming operations where a blow torch is used therefore require some expertise, since no visual signs reveal how close the material is to melting. Aluminium alloys, like all structural alloys, also are subject to internal stresses following heating operations such as welding and casting. The problem with aluminium alloys in this regard is their low melting point, which make them more susceptible to distortions from thermally induced stress relief. Controlled stress relief can be done during manufacturing by heat-treating the parts in an oven, followed by gradual cooling—in effect annealing the stresses.

The low melting point of aluminium alloys has not precluded their use in rocketry; even for use in constructing combustion chambers where gases can reach 3500 K. The Agena upper stage engine used a regeneratively cooled aluminium design for some parts of the nozzle, including the thermally critical throat region.

Another alloy of some value is aluminium bronze (CuAl5).

History

Ancient Greeks and Romans used aluminium salts as dyeing mordants and as astringents for dressing wounds; alum is still used as a styptic. In 1761, Guyton de Morveau suggested calling the base alum *alumine*. In 1808, Humphry Davy identified the existence of a metal base of alum, which he at first termed *alumium* and later *aluminium*.

The metal was first produced in 1825 (in an impure form) by Danish physicist and chemist Hans Christian Ørsted. He reacted anhydrous aluminium chloride with potassium amalgam and yielded a lump of metal looking similar to tin. Friedrich Wöhler was aware of these experiments and cited them, but after redoing the experiments of Ørsted he concluded that this metal was pure potassium. He conducted a similar experiment in 1827 by mixing anhydrous aluminium chloride with potassium and yielded aluminium. Wöhler is generally cr.ed with isolating aluminium (Latin *alumen*, alum), but also Ørsted can be listed as its discoverer. Further, Pierre Berthier discovered aluminium in bauxite ore and successfully extracted it. Frenchman Henri Etienne Sainte-Claire Deville improved Wöhler's method in 1846, and described his improvements in a book in 1859, chief among these being the substitution of sodium for the considerably more expensive potassium. Deville likely also conceived the idea of the electrolysis of aluminium oxide dissolved in cryolite; Charles Martin Hall and Paul Héroult might have developed the more practical process after Deville.

Figure 36: The statue of the Anteros (commonly mistaken for either *The Angel of Christian Charity* or *Eros*) in Piccadilly Circus, London, was made in 1893 and is one of the first statues to be cast in aluminium.

Before the Hall-Héroult process was developed in the late 1880s, aluminium was exceedingly difficult to extract from its various ores.

This made pure aluminium more valuable than gold. Bars of aluminium were exhibited at the Exposition Universelle of 1855. Napoleon III, Emperor of France, is reputed to have given a banquet where the most honoured guests were given aluminium utensils, while the others made do with gold.

Aluminium was selected as the material to be used for the 100 ounce (2.8 kg) capstone of the Washington Monument in 1884, a time when one ounce (30 grams) cost the daily wage of a common worker on the project; The capstone, which was set in place on December 6, 1884, in an elaborate dedication ceremony, was the largest single piece of aluminium cast at the time, when aluminium was as expensive as silver.

The Cowles companies supplied aluminium alloy in quantity in the United States and England using smelters like the furnace of Carl Wilhelm Siemens by 1886. Charles Martin Hall of Ohio in the U.S. and Paul Héroult of France independently developed the Hall-Héroult electrolytic process that made extracting aluminium from minerals cheaper and is now the principal method used worldwide. Hall's process, in 1888 with the financial backing of Alfred E. Hunt, started the Pittsburgh Reduction Company today known as Alcoa. Héroult's process was in production by 1889 in Switzerland at Aluminium Industrie, now Alcan, and at British Aluminium, now Luxfer Group and Alcoa, by 1896 in Scotland.

By 1895, the metal was being used as a building material as far away as Sydney, Australia in the dome of the Chief Secretary's Building.

Many navies have used an aluminium superstructure for their vessels; the 1975 fire aboard USS *Belknap* that gutted her aluminium superstructure, as well as observation of battle damage to British ships during the Falklands War, led to many navies switching to all steel superstructures. The *Arleigh Burke* class was the first such U.S. ship, being constructed entirely of steel.

Aluminium wire was once widely used for domestic electrical wiring. Owing to corrosion-induced failures, a number of fires resulted. This discontinuation thus illustrates one failed application of the otherwise highly useful metal.

In 2008, the price of aluminium peaked at \$1.45/lb in July but dropped to \$0.70/lb by December.

Etymology

Two variants of the metal's name are in current use, *aluminium* and *aluminium* (besides the obsolete *alumium*). The International Union of Pure and Applied Chemistry (IUPAC) adopted *aluminium* as the standard international name for the element in 1990 but, three years later, recognised *aluminium* as an acceptable variant. Hence their periodic table includes both. IUPAC prefers the use of *aluminium* in its internal publications, although nearly as many IUPAC publications use the spelling *aluminium.*

Most countries use the spelling *aluminium.* In the United States, the spelling *aluminium* predominates. The Canadian Oxford Dictionary prefers *aluminium,* whereas the Australian Macquarie Dictionary prefers *aluminium.* In 1926, the American Chemical Society officially decided to use *aluminium* in its publications; American dictionaries typically label the spelling *aluminium* as a British variant.

The name *aluminium* derives from its status as a base of alum. It is borrowed from Old French; its ultimate source, *alumen,* in turn is a Latin word that literally means "bitter salt".

The earliest citation given in the Oxford English Dictionary for any word used as a name for this element is *alumium,* which British chemist and inventor Humphry Davy employed in 1808 for the metal he was trying to isolate electrolytically from the mineral *alumina.* The citation is from the journal *Philosophical Transactions of the Royal Society of London*: "Had I been so fortunate as to have obtained more certain evidences on this subject, and to have procured the metallic substances I was in search of, I should have proposed for them the names of silicium, alumium, zirconium, and glucium."

Davy settled on *aluminium* by the time he published his 1812 book *Chemical Philosophy*: "This substance appears to contain a peculiar metal, but as yet Aluminium has not been obtained in a perfectly free state, though alloys of it with other metalline substances have been procured sufficiently distinct to indicate the probable nature of alumina." But the same year, an anonymous contributor to the *Quarterly Review,* a British political-literary journal, in a review of Davy's book, objected to *aluminium* and proposed the name *aluminium,* "for so we shall take the liberty of writing the word, in preference to aluminium, which has a less classical sound."

The *-ium* suffix conformed to the precedent set in other newly discovered elements of the time: potassium, sodium, magnesium, calcium, and strontium (all of which Davy isolated himself). Nevertheless, *-um* spellings for elements were not unknown at the time, as for example platinum, known to Europeans since the 16th century, molybdenum, discovered in 1778, and tantalum, discovered in 1802. The *-um* suffix is consistent with the universal spelling alumina for the oxide, as lanthana is the oxide of lanthanum, and magnesia, ceria, and thoria are the oxides of magnesium, cerium, and thorium respectively.

The spelling used throughout the 19th century by most U.S. chemists was *aluminium*, but common usage is less clear. The *aluminium* spelling is used in the Webster's Dictionary of 1828. In his advertising handbill for his new electrolytic method of producing the metal 1892, Charles Martin Hall used the *-um* spelling, despite his constant use of the *-ium* spelling in all the patents he filed between 1886 and 1903. It has consequently been suggested that the spelling reflects an easier to pronounce word with one fewer syllable, or that the spelling on the flier was a mistake. Hall's domination of production of the metal ensured that the spelling *aluminium* became the standard in North America; the *Webster Unabridged Dictionary* of 1913, though, continued to use the *-ium* version.

Health Concerns

Despite its natural abundance, aluminium has no known function in biology. It is remarkably nontoxic, aluminium sulfate having an LD50 of 6207 mg/kg (oral, mouse), which corresponds to 500 grams for a 80 kg person. Despite the extremely low acute toxicity, the health effects of aluminium are of interest in view of the widespread occurrence of the element in the environment and in commerce.

Some toxicity can be traced to deposition in bone and the central nervous system, which is particularly increased in patients with reduced renal function. Because aluminium competes with calcium for absorption, increased amounts of dietary aluminium may contribute to the reduced skeletal mineralisation (osteopenia) observed in preterm infants and infants with growth retardation. In very high doses, aluminium can cause neurotoxicity, and is associated with altered function of the blood-brain barrier. A small percentage of people are allergic to aluminium and experience contact dermatitis, digestive disorders, vomiting or other symptoms upon contact or ingestion of products containing aluminium, such as deodorants or antacids. In those without allergies, aluminium is not as toxic as heavy metals, but there is evidence of some toxicity if it is consumed in excessive amounts. Although the use of aluminium cookware has not been shown to lead to aluminium toxicity in general, excessive consumption of antacids containing aluminium compounds and excessive use of aluminium-containing antiperspirants provide more significant exposure levels. Studies have shown that consumption of acidic foods or liquids with aluminium significantly increases aluminium absorption, and maltol has been shown to increase the accumulation of aluminium in nervous and osseus tissue. Furthermore, aluminium increases estrogen-related gene expression in human breast cancer cells cultured in the laboratory. The estrogen-like effects of these salts have led to their classification as a metalloestrogen.

The effects of aluminium in antiperspirants has been examined over the course of decades with little evidence of skin irritation. Nonetheless, its occurrence in antiperspirants, dyes (such as aluminium lake), and food additives is controversial in some quarters. Although there is little evidence that normal exposure to aluminium presents a risk to healthy adults, some studies point to risks associated with increased exposure to the metal. Aluminium in food may be absorbed more than aluminium from water. Some researchers have expressed concerns that the aluminium in antiperspirants may increase the risk of breast cancer, and aluminium has controversially been implicated as a factor in Alzheimer's disease. The Camelford water pollution incident involved a number of people consuming aluminium sulfate. Investigations of the long-term health effects are still ongoing, but elevated brain aluminium concentrations have been found in post-mortem examinations of victims, and further research to determine if there is a link with cerebral amyloid angiopathy has been commissioned.

According to The Alzheimer's Society, the overwhelming medical and scientific opinion is that studies have not convincingly demonstrated a causal relationship between aluminium and Alzheimer's disease. Nevertheless, some studies, such as those on the PAQUID cohort, cite aluminium exposure as a risk factor for Alzheimer's disease. Some brain plaques have been found to contain increased levels of the metal. Research in this area has been inconclusive; aluminium accumulation may be a consequence of the disease rather than a causal agent. In any event, if there is any toxicity of aluminium, it must be via a very specific mechanism, since total human exposure to the element in the form of naturally occurring clay in soil and dust is enormously large over a lifetime. Scientific consensus does not yet exist about whether aluminium exposure could directly increase the risk of Alzheimer's disease.

Effect on Plants

Aluminium is primary among the factors that reduce plant growth on acid soils. Although it is generally harmless to plant growth in pH-neutral soils, the concentration in acid soils of toxic Al^{3+} cations increases and disturbs root growth and function.

Most acid soils are saturated with aluminium rather than hydrogen ions. The acidity of the soil is therefore a result of hydrolysis of aluminium compounds. This concept of "corrected lime potential" to define the degree of base saturation in soils became the basis for procedures now used in soil testing laboratories to determine the "lime requirement" of soils.

Wheat's adaptation to allow aluminium tolerance is such that the aluminium induces a release of organic compounds that bind to the harmful aluminium cations. Sorghum is believed to have the same tolerance mechanism. The first gene for aluminium tolerance has been identified in wheat. It was shown that sorghum's aluminium tolerance is controlled by a single gene, as for wheat. This is not the case in all plants.

Alunite

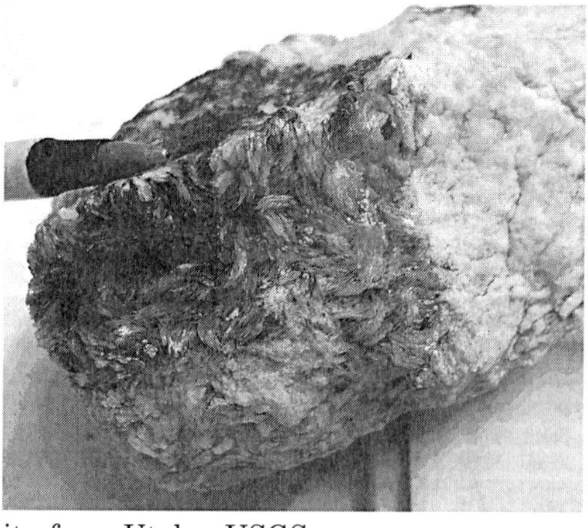

Figure 37: Alunite from Utah - USGS

Alunite is a sulfate mineral that was first observed in the 15th century in Monti della Tolfa, north of Rome, where it was mined for the manufacture of alum. First called *aluminilite* by J.C. Delamétherie in 1797, this name was contracted by François Beudant in 1824 to alunite.

Distinct crystals of alunite are rarely found in cavities in the massive material. Alunite crystallises in the hexagonal system with crystals forming trigonal pyramids that resemble rhombohedra with interfacial angles of 90° 50', so that they resemble cubes in appearance. Minute glistening crystals have also been found loose in cavities in altered rhyolite. Alunite varies in colour from white to yellow gray. The hardness is 4 and the specific gravity is between 2.6 and 2.8. The mineral is a hydrated aluminium potassium sulfate, $KAl_3(SO_4)_2(OH)_6$. Sodium substitutes for potassium and when high in sodium it is called natroalunite. It is insoluble in water or weak acids, but soluble in sulfuric acid. Jarosite is an iron analogue in which Fe^{3+} replaces the aluminium. It occurs as a secondary mineral on iron sulfate ores.

Figure 38: Alunite from Slovakia

Figure 39: Alunite Marysvale, Utah

Alunite occurs as veins and replacement masses in trachyte, rhyolite, and similar potassium rich volcanic rocks. It is formed by the action of sulfuric acid bearing solutions on these rocks during the oxidation and leaching of metal sulfide deposits. Alunite also is found near volcanic fumaroles. The white, finely granular masses closely resemble finely granular limestone, dolomite, anhydrite, and magnesite in appearance. The more compact kinds from Hungary are so hard and tough that they have been used for millstones.

Historically extensive deposits were mined in Tuscany and Hungary, and at Bulahdelah, Australia. Currently found at Tolfa, Italy. In the United States it is found in the San Juan district of Colourado; Goldfield, Nevada; the ghost town of Alunite, Utah near Marysvale; and Red Mountain near Patagonia, Arizona. The Arizona occurrence lies appropriately above a canyon named *Alum Gulch*. Alunite is mined as an ore of both potassium and aluminium at Marysvale. Some of the ore deposits were located by airborne and satellite multispectral imaging.

Alunogen

Figure 40: Alunogen

Alunogen also called feather alum and hair salt is a colourless to white (although often coloured by impurities, like iron substituting for aluminium) fibrous to needle-like aluminium sulfate mineral often found on the walls of mines and quarries as secondary mineral. It can be found in the oxidation zones of some ore deposits as well as on burning coal dumps (i.e. as the product of millosevichite hydratation).

Alunogen has the chemical formula $Al_2(SO_4)_3 \cdot 17H_2O$. More precise, crystallochemical formula, can be written as: $[Al(H_2O)_6]_2(SO_4)_3 \cdot 5H_2O$.

The second formula shows that H_2O in the alunogen formula occurs both as ligand (coordinative form) and loosely-bounded (crystallisation) form. The second form can be easily removed just by slight grinding of the mineral (by preserving the unique structure).

Amblygonite

Figure 41: Amblygonite

Amblygonite is a fluorophosphate mineral, $(Li,Na)AlPO_4(F,OH)$, composed of lithium, sodium, aluminium, phosphate, fluoride and hydroxide. The mineral occurs in pegmatite deposits and is easily mistaken for albite and other feldspars. Its density, cleavage and flame test for lithium are diagnostic. Amblygonite forms a series with *montebrasite*, the low fluorine endmember. Geologic occurrence is in granite pegmatites, high-temperature tin veins, and greisens. Amblygonite occurs with spodumene, apatite, lepidolite, tourmaline, and other lithium-bearing minerals in pegmatite veins. It contains about 10% lithium, and has been utilised as a source of lithium. The chief commercial sources have historically been the deposits of California and France.

History

The mineral was first discovered in Saxony by August Breithaupt in 1817, and named by him from the Greek *amblus*, blunt, and *gouia*, angle, because of the obtuse angle between the cleavages. Later it was found at Montebras, Creuse, France, and at Hebron in Maine; and because of slight differences in optical character and chemical composition the names montebrasite and hebronite have been applied to the mineral from these localities. It has been discovered in considerable quantity at Pala in San Diego county, California; Caceres, Spain; and the Black Hills of South Dakota. The largest documented single crystal of amblygonite measured 7.62x2.44x1.83 m^3 and weighed ~102 tons.

Gemology

Transparent amblygonite has been faceted and used as a gemstone. As a gemstone set into jewelry it is vulnerable to breakage and abrasion from general wear, as its hardness and toughness are poor. The main sources for gem material are Brazil and the U.S.. Australia, France, Germany, Namibia, Norway, and Spain have also produced gem quality amblygonite.

Figure 42: Amblygonite from Taquaral, Itinga, Minas Gerais, Brazil. Scale at bottom is one inch, with a rule at one cm.

Ameghinite

Ameghinite ($Na[H_4B_3O_7]$ or $NaB_3O_3(OH)_4$) is a mineral found in Argentina. It is a soft mineral with a Mohs hardness of 2-3. Ameghinite has a Monoclinic crystal system.

It was first described in 1967 for an occurrence in the Tincalayu Mine, Salar Del Hombre Muerto, Salta, Argentina. It was named for Argentine geologist brothers, Carlos Ameghino (1865–1936) and Florentino Ameghino (1854–1911).

Amphibole

Figure 43: Amphibole (Hornblende)

Amphibole is the name of an important group of generally dark-coloured rock-forming inosilicate minerals, composed of double chain SiO_4 tetrahedra, linked at the vertices and generally containing ions of iron and/or magnesium in their structures.

Mineralogy

Amphiboles crystallise into two crystal systems, monoclinic and orthorhombic. In chemical composition and general characteristics they are similar to the pyroxenes. The chief differences from pyroxenes are that (i) amphiboles contain essential hydroxyl (OH) or halogen (F, Cl) and (ii) the basic structure is a double chain of tetrahedra (as opposed to the single chain structure of pyroxene). Most apparent, in hand specimens, is that amphiboles form oblique cleavage planes (at around 120 degrees), whereas pyroxenes have cleavage angles of approximately 90 degrees. Amphiboles are also specifically less dense than the corresponding pyroxenes. In optical characteristics, many amphiboles are distinguished by their stronger pleochroism and by the smaller angle of extinction (Z angle c) on the plane of symmetry. Amphiboles are the primary constituent of amphibolites.

In Rocks

Amphiboles are minerals of either igneous or metamorphic origin; in the former case occurring as constituents (hornblende) of igneous rocks, such as granite, diorite, andesite and others. Calcium is sometimes a constituent of naturally occurring amphiboles.(C. Michael Hogan. 2010) Those of metamorphic origin include examples such as those developed in limestones by contact metamorphism (tremolite) and those formed by the alteration of other ferromagnesian minerals (hornblende). Pseudomorphs of amphibole after pyroxene are known as uralite.

History and Etymology

The name amphibole was used by René Just Haüy to include tremolite, actinolite, tourmaline and hornblende. The group was so named by Haüy in allusion to the protean variety, in composition and appearance, assumed by its minerals. This term has since been applied to the whole group. Numerous sub-species and varieties are distinguished, the more important of which are tabulated below in two series. The formulae of each will be seen to be built on the general double-chain silicate formula RSi_4O_{11}.

Mineral Species

Chemical Formulae

Orthorhombic series:

- Anthophyllite $(Mg,Fe)_7Si_8O_{22}(OH)_2$
- Holmquistite $Li_2Mg_3Al_2Si_8O_{22}(OH)_2$

Monoclinic series:

- Tremolite $Ca_2Mg_5Si_8O_{22}(OH)_2$
- Actinolite $Ca_2(Mg,Fe)_5Si_8O_{22}(OH)_2$
- Cummingtonite $Fe_2Mg_5Si_8O_{22}(OH)_2$
- Grunerite $Fe_7Si_8O_{22}(OH)_2$
- Hornblende $Ca_2(Mg,Fe,Al)_5(Al,Si)_8O_{22}(OH)_2$
- Glaucophane $Na_2(Mg,Fe)_3Al_2Si_8O_{22}(OH)_2$
- Riebeckite (or Crocidolite) $Na_2Fe^{2+}{}_3Fe^{3+}{}_2Si_8O_{22}(OH)_2$
- Arfvedsonite $Na_3Fe^{2+}{}_4Fe^{3+}Si_8O_{22}(OH)_2$
- Richterite $Na_2Ca(Mg,Fe)_5Si_8O_{22}(OH)_2$
- Pargasite $NaCa_2Mg_3Fe^{2+}Si_6Al_3O_{22}(OH)_2$
- Winchite $(CaNa)Mg_4(Al,Fe^{3+})Si_8O_{22}(OH)_2$

Descriptions

On account of the wide variations in chemical composition, the different members vary considerably in properties and general appearance.

Anthophyllite occurs as brownish, fibrous or lamellar masses with hornblende in mica-schist at Kongsberg in Norway and some other localities. An aluminous related species is known as gedrite and a deep green Russian variety containing little iron as kupfferite.

Hornblende is an important constituent of many igneous rocks. It is also an important constituent of amphibolites formed by metamorphism of basalt.

Actinolite is an important and common member of the monoclinic series, forming radiating groups of acicular crystals of a bright green or greyish-green colour. It occurs frequently as a constituent of greenschists. The name is a translation of the old German word *Strahlstein* (radiated stone).

Glaucophane, crocidolite, riebeckite and arfvedsonite form a somewhat special group of alkali-amphiboles. The first two are blue fibrous minerals, with glaucophane occurring in blueschists and crocidolite (blue asbestos) in ironstone formations, both resulting from dynamo-metamorphic processes. The latter two are dark green minerals, which occur as original constituents of igneous rocks rich in sodium, such as nepheline-syenite and phonolite.

Pargasite is a rare magnesium-rich amphibole with essential sodium, usually found in ultramafic rocks. For instance, it occurs in uncommon mantle xenoliths, carried up by kimberlite. It is hard, dense, black and usually idiomorphic, with a red-.

Analcime

Figure 44: Analcime

Analcime or analcite is a white, grey, or colourless tectosilicate mineral. Analcime consists of hydrated sodium aluminium silicate in cubic crystalline form. Its chemical formula is $NaAlSi_2O_6 \cdot H_2O$. Minor amounts of potassium and calcium substitute for sodium. A silver-bearing synthetic variety also exists (Ag-analcite).

Analcime is usually classified as a zeolite mineral, but structurally and chemically it is more similar to the feldspathoids. Analcime occurs as a primary mineral in analcime basalt and other alkaline igneous rocks. It also occurs as cavity and vesicle fillings associated with prehnite, calcite, and zeolites.

Locations include the Cyclopean Islands east off Sicily and near Trentino in northern Italy; Victoria in Australia; Kerguelen Island in the Indian Ocean; in the Lake Superior copper district of Michigan, Bergen Hill, New Jersey, Golden, Colourado, and at Searles Lake, California in the United States; and at Cape Blomidon, Nova Scotia and Mont Saint-Hilaire, Quebec in Canada; and in Iceland.

Anapaite

Figure 44: Anapaite from the Ukraine

Anapaite is a calcium iron phosphate mineral, $Ca_2Fe^{2+}(PO)_4]_2 \cdot 4H_2O$. It is a mineral that typically occurs in cavities in fossil bearing sedimentary rocks. It is also found in phosphate bearing iron ores.

The spectacular environment that anapaite usually grows in makes it a very popular mineral for collectors. Named after the type locality Anapa, Taman Peninsula, Russia. Some of the best localities are Kertch (Crimea, Ukraine) and Bellaver de Cerdena (Gerona, Spain).

Anatase

Anatase is one of the three mineral forms of titanium dioxide, the other two being brookite and rutile. It is always found as small, isolated and sharply developed crystals, and like rutile, a more commonly occurring modification of titanium dioxide, it crystallises in the tetragonal system; but, although the degree of symmetry is the same for both, there is no relation between the interfacial angles of the two minerals, except in the prism-zone of 45° and 90°. The common pyramid of anatase, parallel to the faces of which there are perfect cleavages, has an angle over the polar edge of 82°9', the corresponding angle of rutile being 56°52½'. It was on account of this steeper pyramid of anatase that the mineral was named, by René Just Haüy in 1801, from the Greek *anatasis*, "extension", the vertical axis of the crystals being longer than in rutile. There are also important differences between the physical characters of anatase and rutile: the former is less hard (5.5–6 vs. 6-6.5 Mohs) and dense (specific gravity about 3.9 vs. 4.2). Also, anatase is optically negative whereas rutile is positive, and its luster is even more strongly adamantine or metallic-adamantine than that of rutile.

Crystal Habit

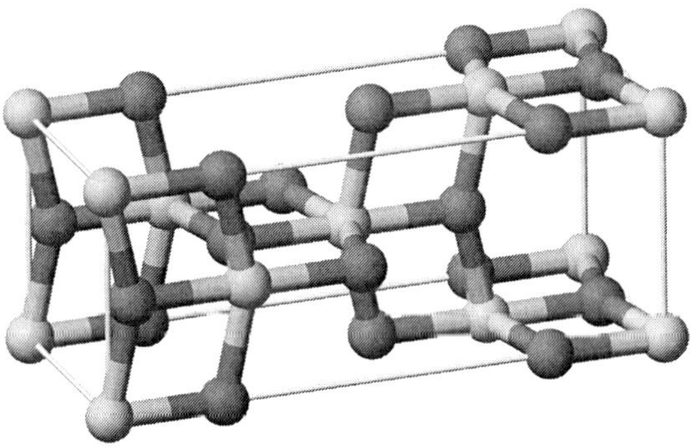

Figure 45: Crystal structure of anatase

Two types or habits of anatase crystals may be distinguished. The commoner occurs as simple acute double pyramids with an indigo-blue to black colour and steely luster. Crystals of this kind are abundant at Le Bourg-d'Oisans in Dauphiné, where they are associated with rock-crystal, feldspar, and axinite in crevices in granite and mica-schist. Similar crystals, but of microscopic size, are widely distributed in sedimentary rocks, such as sandstones, clays, and slates, from which they may be separated by washing away the lighter constituents of the powdered rock. Crystals of the second type have numerous pyramidal faces developed, and they are usually flatter or sometimes prismatic in habit; the colour is honey-yellow to brown. Such crystals closely resemble xenotime in appearance and, indeed, were for a long time supposed to belong to this species, the special name *wiserine* being applied to them. They occur attached to the walls of crevices in the gneisses of the Alps, the Binnenthal near Brig in canton Valais, Switzerland, being a well-known locality. Naturally occurring pseudomorphs of rutile after anatase are also known.

While anatase is not an equilibrium phase of TiO_2, it is kinetically stabilised. At temperatures between 550 and about 1000 °C, anatase transforms to the equilibrium rutile phase, increasing its specific gravity to 4.2. The temperature of this transformation strongly depends on the impurities or dopants present in the material as well as on the morphology of the sample.

Synthetic Anatase

Crystals of anatase can be prepared in laboratories by chemical methods such as sol-gel processing. Examples include controlled hydrolysis of titanium tetrachloride ($TiCl_4$) or titanium alkoxides. Such synthetic anatases are of considerable interest for photocatalysis and for Graetzel-cell photovoltaics.

Alternate and Obsolete Names

Another name commonly in use for this mineral is *octahedrite*, a name which, indeed, is earlier than anatase, and given because of the common (acute) octahedral habit of the crystals. Other names, now obsolete, are oisanite and dauphinite, from the well-known French locality.

Andalusite

Andalusite is an aluminium nesosilicate mineral with the chemical formula Al_2SiO_5.

The variety chiastolite commonly contains dark inclusions of carbon or clay which form a checker-board pattern when shown in cross-section.

A clear variety first found in Andalusia, Spain can be cut into an interesting gemstone. Faceted andalusite stones give a play of red, green, and yellow colours that resembles a muted form of iridescence, although the colours are actually the result of unusually strong pleochroism.

Figure 46: Andalusite, Tyrol Austria.

It is associated with mica schist which increases alkali content in ultimate product and so it has not been exploited economically so far.

Occurrence

Andalusite is a common regional metamorphic mineral which forms under low pressure and moderate to high temperatures. Called Lapis Crucifer in ancient texts. The minerals kyanite and sillimanite are polymorphs of andalusite, each occurring under different temperature-pressure regimes and are therefore rarely found together in the same rock. Because of this the three minerals are a useful tool to help identify the pressure-temperature paths of the host rock in which they are found.

It was first described and named after the type locality in the Ronda Massif, Málaga, Andalusia, Spain in 1789.

Andesine

Andesine is a silicate mineral, a member of the plagioclase feldspar solid solution series. Its chemical formula is (Ca, Na)(Al, Si)$_4$O$_8$, where Ca/(Ca + Na) (% Anorthite) is between 30%-50%. The formula may be written as $Na_{0.7\text{-}0.5}Ca_{0.3\text{-}0.5}Al_{1.3\text{-}1.5}Si_{2.7\text{-}2.5}O_8$. The plagioclase feldspars are a continuous solid solution series and as such the accurate identification of individual members requires detailed optical study, chemical analysis or density measurements.

Figure 47: Andesine

Refractive indices and specific gravity increase directly with calcium content.

Name and Discovery

Andesine was first described in 1841 for an occurrence in the Marmato mine, Marmato, Cauca, Chocó Department, Colombia. The name is for the Andes due to its abundance in the andesite lavas in those mountains.

Occurrence

Andesine occurs in intermediate igneous rocks such as diorite, syenite and andesite. It characteristically occurs in metamorphic rocks of granulite to amphibolite facies commonly exhibiting antiperthite texture. It also occurs as detrital grains in sedimentary rocks. It is commonly associated with quartz, potassium feldspar, biotite, hornblende and magnetite.

Anglesite

Figure 48: Anglesite Touizit Morroco

Anglesite is a lead sulfate mineral with the chemical formula $PbSO_4$. It occurs as an oxidation product of primary lead sulfide ore, galena. Anglesite occurs as prismatic orthorhombic crystals and earthy masses, and is isomorphous with barite and celestine. It contains 74% of lead by mass and therefore has a high specific gravity of 6.3. Colour is white or gray with pale yellow streaks. It may be dark gray if impure.

Figure 49: Anglesite crystal from Touissit District, Morocco (size: 2.8×1.6×0.5 cm)

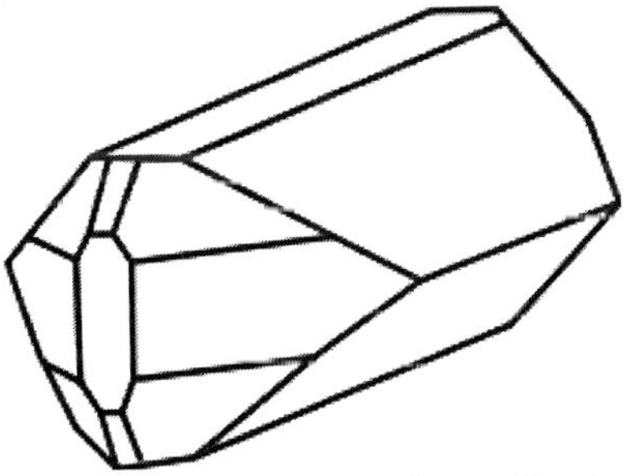

Figure 50: Anglesite diagram illustrating its orthorhombic crystalline form

It was first recognised as a mineral species by William Withering in 1783, who discovered it in the Parys copper-mine in Anglesey; the name anglesite, from this locality, was given by F. S. Beudant in 1832. The crystals from Anglesey, which were formerly found abundantly on a matrix of dull limonite, are small in size and simple in form, being usually bounded by four faces of a prism and four faces of a dome; they are brownish-yellow in colour owing to a stain of limonite. Crystals from some other localities, notably from Monteponi in Sardinia, are transparent and colourless, possessed of a brilliant adamantine lustre, and usually modified by numerous bright faces.

The variety of combinations and habits presented by the crystals is very extensive, nearly two hundred distinct forms being figured by V. von Lang in his monograph of the species; without measurement of the angles the crystals are frequently difficult to decipher. There are distinct cleavages parallel to the faces of the prism (110) and the basal plane (001), but these are not so well developed as in the isomorphous minerals barite and celestite.

Anglesite is a mineral of secondary origin, having been formed by the oxidation of galena in the upper parts of mineral lodes where these have been affected by weathering processes.

At Monteponi the crystals encrust cavities in glistening granular galena; and from Leadhills, in Scotland, pseudomorphs of anglesite after galena are known. At most localities it is found as isolated crystals in the lead-bearing lodes, but at some places, in Australia and Mexico, it occurs as large masses, and is then mined as an ore of lead.

Anhydrite

Figure 51: Anhydrite, Chihuahua, Mexico

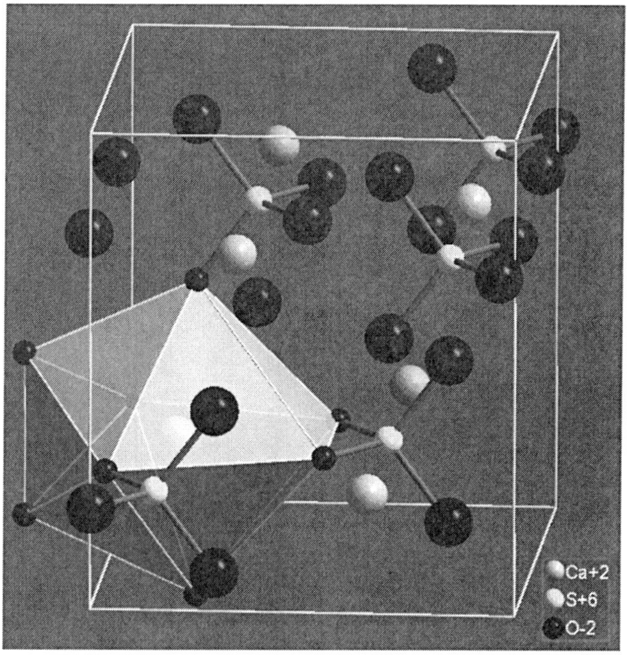

Figure 52: Crystal structure of anhydrite

Anhydrite is a mineral – anhydrous calcium sulfate, $CaSO_4$. It is in the orthorhombic crystal system, with three directions of perfect cleavage parallel to the three planes of symmetry.

It is not isomorphous with the orthorhombic barium (baryte) and strontium (celestine) sulfates, as might be expected from the chemical formulas. Distinctly developed crystals are somewhat rare, the mineral usually presenting the form of cleavage masses.

The hardness is 3.5 and the specific gravity 2.9. The colour is white, sometimes greyish, bluish or purple. On the best developed of the three cleavages the lustre is pearly, on other surfaces it is vitreous.

When exposed to water, anhydrite readily transforms to the more commonly occurring gypsum, ($CaSO_4 \cdot 2H_2O$) by the absorption of water. This transformation is reversible, with gypsum or calcium sulfate hemihydrate forming anhydrite by heating to ~200°C under normal atmospheric conditions.

Anhydrite is commonly associated with calcite, halite, and sulfides such as galena, chalcopyrite, molybdenite and pyrite in vein deposits.

Occurrence

Anhydrite is most frequently found in evaporite deposits with gypsum; it was, for instance, first discovered, in 1794, in a salt mine near Hall in Tirol. In this occurrence depth is critical since nearer the surface anhydrite has been altered to gypsum by absorption of circulating ground water.

From an aqueous solution calcium sulfate is deposited as crystals of gypsum, but when the solution contains an excess of sodium or potassium chloride anhydrite is deposited if temperature is above 40 °C. This is one of the several methods by which the mineral has been prepared artificially, and is identical with its mode of origin in nature, the mineral is common in salt basins.

Tidal Flat Nodules

Anhydrite occurs in a tidal flat environment in the Persian Gulf sabkhas as massive diagenetic replacement nodules. Cross sections of these nodular masses have a *netted* appearance and have been referred to as *chicken wire anhydrite*. Nodular anhydrite occurs as replacement of gypsum in a variety of sedimentary depositional environments.

Salt Dome Cap Rocks

Massive amounts of anhydrite occur when salt domes form a caprock. Anhydrite is 1–3% of the salt in salt domes and is generally left as a cap at the top of the salt when the halite is removed by pore waters. The typical cap rock is a salt, topped by a layer of anhydrite, topped by patches of gypsum, topped by a layer of calcite. Interaction with oil can reduce SO_4 creating calcite, water, and hydrogen sulfide (H_2S).

Naming History

The name anhydrite was given by A. G. Werner in 1804, because of the absence of water of crystallisation, as contrasted with the presence of water in gypsum. Some obsolete names for the species are muriacite and karstenite; the former, an earlier name, being given under the impression that the substance was a chloride (muriate). A peculiar variety occurring as contorted concretionary masses is known as tripe-stone, and a scaly granular variety, from Volpino, near Bergamo, in Lombardy, as vulpinite; the latter is cut and polished for ornamental purposes.

Ankerite

Figure 53: Ankerite on quartz from Peru.

Ankerite is a calcium, iron, magnesium, manganese carbonate mineral of the group of rhombohedral carbonates with formula: Ca(Fe, Mg, Mn)$(CO_3)_2$. In composition it is closely related to dolomite, but differs from this in having magnesium replaced by varying amounts of iron(II) and manganese.

The crystallographic and physical characters resemble those of dolomite and siderite. The angle between the perfect rhombohedral cleavages is 73° 48', the hardness is 3.5 to 4, and the specific gravity is 2.9 to 3.1. The colour is white, grey or reddish to yellowish brown.

Ankerite occurs with siderite in deposits of iron-ore. It is one of the minerals of the dolomite-siderite series, to which the terms brown-spar, pearl-spar and bitter-spar have been historically loosely applied.

Ankerite can result from hydrothermal or direct groundwater precipitation. It can also be the result of metamorphic recrystallisation of iron-rich sedimentary rocks. It is often found as a gangue mineral associated with gold and a variety of sulfide minerals in ore deposits.

It was first recognised as a distinct species by W. von Haidinger in 1825, and named by him after Matthias Joseph Anker (1771-1843) of Styria, an Austrian mineralogist.

Annabergite

Annabergite is an arsenate mineral consisting of a hydrous nickel arsenate, $Ni_3(AsO_4)_2 \cdot 8H_2O$, crystallising in the monoclinic system and isomorphous with vivianite and erythrite. Crystals are minute and capillary and rarely met with, the mineral occurring usually as soft earthy masses and encrustations. A fine apple-green colour is its characteristic feature.

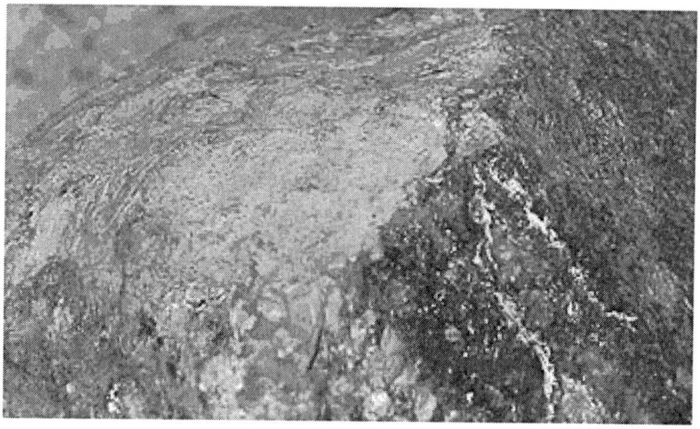

Figure 54: Annabergite, Dobšiná, Slovakia

It was long known (since 1758) under the name nickel ochre; the name annabergite was proposed by H. J. Brooke and W H. Miller in 1852, from Annaberg in Saxony, one of the localities of the mineral. It occurs with ores of nickel, of which it is a product of alteration. A variety, from Creetown in Kirkcudbrightshire, in which a portion of the nickel is replaced by calcium, has been called *dudgeonite*, after P. Dudgeon, who found it.

Figure 55: Annabergite from Lavrion (Laurium), Greece

Closely related is *cabrerite* wherein some of the nickel is replaced by magnesium. It is named for Sierra Cabrera in Spain where it was originally found.

Anorthite

Figure 56: Anorthite crystals in a basalt vug from Vesuvius (size:6.9 x 4.1 x 3.8 cm)]]

Anorthite is the calcium endmember of plagioclase feldspar. Plagioclase is an abundant mineral in the Earth's crust. The formula of pure anorthite is $CaAl_2Si_2O_8$.

Mineralogy

Figure 57: Anorthite crystal in lava from Miyake Island, Japan (size: 2.4 x 1.7 x 1.7 cm)

Anorthite is the calcium-rich endmember of the plagioclase solid solution series, the other endmember being albite, $NaAlSi_3O_8$. Anorthite also refers to plagioclase compositions with more than 90 molecular percent of the anorthite endmember.

Occurrence

Anorthite is a rare compositional variety of plagioclase. It occurs in mafic igneous rock. It also occurs in granulite facies metamorphic rocks, in metamorphosed carbonate rocks and corundum deposits. Its type localities are Monte Somma and Valle di Fassa, Italy. It was first described in 1823. It is more rare in surficial rocks than it normally would be due to its high weathering potential in the Goldich dissolution series.

It also makes up much of the lunar highlands; the Genesis Rock is made of anorthosite, which is composed largely of anorthite. Anorthite was discovered in samples from comet Wild 2, and the mineral is an important constituent of Ca-Al-rich inclusions in rare varieties of chondritic meteorites.

Anorthoclase

Figure 58: An Anorthoclase crystal from Mount Erebus, Antarctica.

The mineral anorthoclase ($(Na,K)AlSi_3O_8$) is a crystalline solid solution in the alkali feldspar series, in which the sodium-aluminium silicate member exists in larger proportion. It typically consists of between 10 and 36 percent of $KAlSi_3O_8$ and between 64 and 90 percent of $NaAlSi_3O_8$.

Anthophyllite

Figure 59: Anthophyllite

Anthophyllite is an amphibole mineral: $(Mg, Fe)_7Si_8O_{22}(OH)_2$, magnesium iron inosilicate hydroxide. Anthophyllite is polymorphic with cummingtonite. Some forms of anthophyllite are lamellar or fibrous and are used as asbestos. The name is derived from the Latin word *anthophyllum*, meaning *clove*, an allusion to the most common colour of the mineral.

Occurrence

Anthophylite is the product of metamorphism of magnesium-rich rocks, especially ultrabasic igneous rocks and impure dolomitic shales. It also forms as a retrograde product rimming relict orthopyroxenes and olivine, and as an accessory mineral in cordierite-bearing gneisses and schists.

Anthophyllite also occurs as a retrograde metamorphic mineral derived from ultramafic rocks along with serpentinite. Geographically, it occurs in Pennsylvania, southwestern New Hampshire, central Massachusetts, Franklin, North Carolina, and in the Gravelly Range and Tobacco Root Mountains of southwest Montana.

Occurrence in Ultramafic Rocks

Anthophyllite is formed by the breakdown of talc in ultramafic rocks in the presence of water and carbon dioxide as a prograde metamorphic reaction. The partial pressure of carbon dioxide (XCO_2) in aqueous solution favours production of anthophyllite. Higher partial pressures of CO_2 reduces the temperature of the *anthophyllite-in* isograd.

Ultramafic rocks in purely hydrous, CO_2-free environments will tend to form serpentinite-antigorite-brucite-tremolite assemblages (dependent on MgO content) or at amphibolite to granulite metamorphic grade, metamorphic pyroxene or olivine. Thus, metamorphic assemblages of ultramafic rocks containing anthophyllite are indicative of at least greenschist facies metamorphism in the presence of carbon dioxide bearing metamorphic fluids.

The typical metamorphic assemblage reactions for low-magnesian (<25% MgO) and high-magnesian (>25% MgO) ultramafic rocks are;

- Olivine + Tremolite + Talc → Olivine + Tremolite + Anthophyllite (low MgO, >550°C, XCO2 <0.6)
- Talc + Tremolite + Magnesite → Tremolite + Anthophyllite + Magnesite (High MgO, >500°C, XCO2 >0.6)
- Talc + Magnesite + Tremolite → Anthophyllite + Tremolite + Magnesite (Low MgO, >500°C, XCO2 >0.6)

Figure 60: Anthophyllite in serpentinised komatiite, Maggie Hays Ni Mine, Western Australia

Retrogressive anthophyllite is relatively rare in ultramafic rocks and is usually poorly developed due to the lower energy state available for metamorphic reactions to progress and also the general dehydration of rock masses during metamorphism. Similarly, the need for substantial components of carbon dioxide in metamorphic fluid restricts the appearance of anthophyllite as a retrograde mineral. The usual metamorphic assemblage of retrograde-altered ultramafic rocks is thus usually a serpentinite or talc-magnesite assemblage.

Retrograde anthophyllite is present most usually in shear zones where fracturing and shearing of the rocks provides a conduit for carbonated fluids during retrogression.

Serpentine Group

Figure 61: Serpentine

Figure 62: Dish of serpentine with inlaid gold fish, 1st century BCE or CE, with 9th century mounts

Figure 63: Necklace and earring set made from semiprecious stones. The circular green beads are Russian serpentine. Also used are jasper (red) and fluorite (blue).

The serpentine group describes a group of common rock-forming hydrous magnesium iron phyllosilicate ($(Mg, Fe)_3Si_2O_5(OH)_4$) minerals; they may contain minor amounts of other elements including chromium, manganese, cobalt and nickel. In mineralogy and gemology, serpentine may refer to any of 20 varieties belonging to the serpentine group. Owing to admixture, these varieties are not always easy to individualize, and distinctions are not usually made. There are three important mineral polymorphs of serpentine: antigorite, chrysotile and lizardite.

Many types of serpentine have been used for jewellery and hardstone carving, sometimes under the name *false jade* or *Teton jade*.

Overview

"Their olive green colour and smooth or scaly appearance is the basis of the name from the Latin *serpentinus*, meaning "serpent rock," according to Best (2003). They have their origins in metamorphic alterations of peridotite and pyroxene. Serpentines may also pseudomorphously replace other magnesium silicates. Alterations may be incomplete, causing physical properties of serpentines to vary widely. Where they form a significant part of the land surface, the soil is unusually high in clay.

Antigorite is the polymorph of serpentine that most commonly forms during metamorphism of wet ultramafic rocks and is stable at the highest temperatures—to over 600 °C at depths of 60 km or so. In contrast, lizardite and chrysotile typically form near the Earth's surface and break down at relatively low temperatures, probably well below 400 °C. It has been suggested that chrysotile is never stable relative to either of the other two serpentine polymorphs.

Samples of the oceanic crust and uppermost mantle from ocean basins document that ultramafic rocks there commonly contain abundant serpentine. Antigorite contains water in its structure, about 13 percent by weight. Hence, antigorite may play an important role in the transport of water into the earth in subduction zones and in the subsequent release of water to create magmas in island arcs, and some of the water may be carried to yet greater depths.

Soils derived from serpentine are toxic to many plants, because of high levels of nickel, chromium, and cobalt; growth of many plants is also inhibited by low levels of potassium and phosphorus and a low ratio of calcium/magnesium. The flora is generally very distinctive, with specialised, slow-growing species. Areas of serpentine-derived soil will show as strips of shrubland and open, scattered small trees (often conifers) within otherwise forested areas; these areas are called serpentine barrens.

Most serpentines are opaque to translucent, light (specific gravity between 2.2 2.9), soft (hardness 2.5–4), infusible and susceptible to acids. All are microcrystalline and massive in habit, never being found as single crystals. Luster may be vitreous, greasy or silky. Colours range from white to grey, yellow to green, and brown to black, and are often splotchy or veined. Many are intergrown with other minerals, such as calcite and dolomite. Occurrence is worldwide; New Caledonia, Canada (Quebec), USA (northern California, Rhode Island, Connecticut, Massachusetts, Maryland and southern Pennsylvania), Afghanistan, Britain (Cornwall and Ireland), Greece (Thessaly), China, France, Korea, Austria (Styria and Carinthia), India (Assam, and Manipur), Myanmar (Burma), New Zealand, Norway and Italy are notable localities.

Rock composed primarily of these minerals is called serpentinite. Serpentines find use in industry for a number of purposes, such as railway ballasts, building materials, and the asbestiform types find use as thermal and electrical insulation (chrysotile asbestos). The asbestos content can be released to the air when serpentine is excavated and if it is used as a road surface, forming a long term health hazard by breathing. Asbestos from serpentine can also appear at low levels in water supplies through normal weathering processes, but there is as yet no identified health hazard associated with use or ingestion. In its natural state, some forms of serpentine react with carbon dioxide and re-release oxygen into the atmosphere.

The more attractive and durable varieties (all of antigorite) are termed "noble" or "precious" serpentine and are used extensively as gems and in ornamental carvings. The town of Bhera in the historic Punjab province of the Indian subcontinent was known for centuries for finishing a relatively pure form of green serpentine obtained from quarries in Afghanistan into lapidary work, cups, ornamental sword hilts, and dagger handles. This high-grade serpentine ore was known as *sang-i-yashm* or to the English, *false jade*, and was used for generations by Indian craftsmen for lapidary work. It is easily carved, taking a good polish, and is said to have a pleasingly greasy feel. Less valuable serpentine ores of varying hardness and clarity are also sometimes dyed to imitate jade. Misleading synonyms for this material include "Suzhou jade", "Styrian jade", and "New jade".

New Caledonian serpentine is particularly rich in nickel, and is the source of most of the world's nickel ore. The Mâori of New Zealand once carved beautiful objects from local serpentine, which they called *tangiwai*, meaning "tears".

The *lapis atracius* of the Romans, now known as verde antique, or verde antico, is a serpentinite breccia popular as a decorative facing stone. In classical times it was mined at Casambala, Thessaly, Greece. Serpentinite marbles are also widely used: Green Connemara marble (or Irish green marble) from Connemara, Ireland (and many other sources), and red Rosso di Levanto marble from Italy. Use is limited to indoor settings as serpentinites do not weather well.

Antigorite

Lamellated antigorite occurs in tough, pleated masses. It is usually dark green in colour, but may also be yellowish, gray, brown or black.

It has a hardness of 3.5–4 and its lustre is greasy. The monoclinic crystals show micaceous cleavage and fuse with difficulty. Antigorite is named after its type locality, the Valle di Antigorio in Italy.

Bowenite is an especially hard serpentine (5.5) of a light to dark apple green colour, often mottled with cloudy white patches and darker veining. It is the serpentine most frequently encountered in carving and jewellery. The name retinalite is sometimes applied to yellow bowenite. The New Zealand material is called tangiwai.

Although not an official species, bowenite is the state mineral of Rhode Island: this is also the variety's type locality. A bowenite cabochon featured as part of the "Our Mineral Heritage Brooch", was presented to First Lady Mrs. Lady Bird Johnson in 1967.

Williamsite is a local varietal name for antigorite that is oil-green with black crystals of chromite or magnetite often included. Somewhat resembling fine jade, williamsite is cut into cabochons and beads. It is found mainly in Maryland and Pennsylvania, USA.

Gymnite is an amorphous form of antigorite. It was originally found in the Bare Hills, Maryland, and is named from the Greek, gymnos meaning bare or naked.

Lizardite

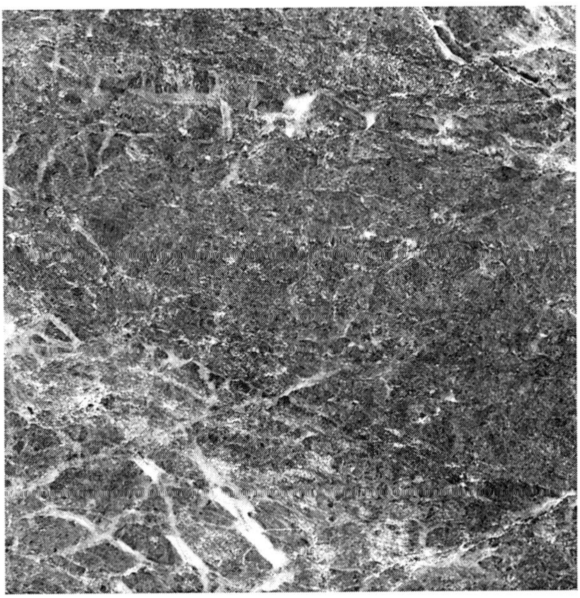

Figure 64: Polished serpentine, from the Val d'Aosta, sold as Gressoney. Used at the United Nations building in New York.

Extremely fine-grained, scaly lizardite (also called orthoantigorite) comprises much of the serpentine present in serpentine marbles. It is triclinic, has one direction of perfect cleavage, and may be white, yellow or green. Lizardite is translucent, soft (hardness 2.5) and has an average specific gravity of 2.57.

It can be pseudomorphous after enstatite, olivine or pyroxene, in which case the name bastite is sometimes applied. Bastite may have a silky lustre.

Lizardite is named after its type locality on the Lizard Peninsula, Cornwall, UK. It is worked by local artisans into various trinkets which are sold to tourists.

Antimony

Figure 65: Silvery lustrous gray

Antimony is a toxic chemical element with the symbol Sb and an atomic number of 51. A lustrous grey metalloid, it is found in nature mainly as the sulfide mineral stibnite (Sb_2S_3). Although the use of antimony is limited by its toxicity, its compounds have been of fundamental value in chemistry – a prominent example being the development of superacids derived from antimony pentafluoride.

Antimony compounds are prominent fire retardants found in many commercial and domestic products. Certain alloys are valuable for use in solders and ball bearings. An emerging application is the use of antimony in microelectronics.

History

Antimony's sulfide compound, antimony (III) sulfide, Sb_2S_3 was recognised in antiquity, at least as early as 3000 BC.

An artifact, said to be part of a vase, made of antimony dating to about 3000 BC was found at Tello, Chaldea (part of present-day Iraq), and a copper object plated with antimony dating between 2500 BC and 2200 BC has been found in Egypt. One contemporary (Austen, at a lecture by Herbert Gladstone, published in 1892) was reported to comment that "we only know of antimony at the present day as a highly brittle and crystalline metal, which could hardly be fashioned into a useful vase, and therefore this remarkable 'find' must represent the lost art of rendering antimony malleable."

However, Moorey was unconvinced that the artefact was indeed a vase, mentioning that Selimkhanov, after his analysis of the Telloh object (published in 1975), "attempted to relate the metal to Transcaucasian natural antimony" (i.e. native metal) and that "the antimony objects from Transcaucasia are all small personal ornaments." This weakens the evidence for a lost art "of rendering antimony malleable."

The first European description of a procedure for isolating antimony is in the book *De la pirotechnia* of 1540 by Vannoccio Biringuccio. This book predates the more famous 1556 book by Agricola, *De re metallica*, even though Agricola has been often incorrectly cr.ed with the discovery of metallic antimony. A text describing the preparation of metallic antimony that was published in Germany in 1604 purported to date from the early fifteenth century, and if authentic it would predate Biringuccio. The book, written in Latin, was called "Currus Triumphalis Antimonii" (The Triumphal Chariot of Antimony), and its putative author was a certain Benedictine monk, writing under the name Basilius Valentinus. An English translation of the "Currus Triumphalis" appeared in English in 1660, under the title The Triumphant Chariot of Antimony. The work remains of great interest, chiefly because it documents how followers of the renegade German physician, Philippus Theophrastus Paracelsus von Hohenheim (of whom Thölde was one), came to associate the practice of alchemy with the preparation of chemical medicines.

Pure antimony was well known to Jâbir ibn Hayyân, sometimes called "the Father of Chemistry", in the 8th century. Here there is still an open controversy: Marcellin Berthelot, who translated a number of Jâbir's books, stated that antimony is never mentioned in them, but other authors claim that Berthelot translated only some of the less important books, while the more interesting ones (some of which might describe antimony) are not yet translated, and their content is completely unknown.

The first natural occurrence of pure antimony ('native antimony') in the Earth's crust was described by the Swedish scientist and local mine district engineer Anton von Swab in 1783. The type-sample was collected from the Sala Silver Mine in the Bergslagen mining district of Sala, Västmanland, Sweden.

Etymology

The ancient words for antimony mostly have, as their chief meaning, kohl, the sulfide of antimony. Pliny the Elder, however, distinguishes between male and female forms of antimony; his male form is probably the sulfide, while the female form, which is superior, heavier, and less friable, is probably native metallic antimony.

The use of Sb as the standard chemical symbol for antimony is due to the 18th century chemical pioneer, Jöns Jakob Berzelius, who used this abbreviation of the name *stibium*. The medieval Latin form, from which the modern languages and late Byzantine Greek, take their names, is *antimonium*.

The origin of this is uncertain; all suggestions have some difficulty either of form or interpretation. The popular etymology, from *íôßìïíá÷üò anti-monachos* or French *antimoine*, still has adherents; this would mean "monk-killer", and is explained by many early alchemists being monks, and antimony being poisonous.

Lippmann conjectured a Greek word *anthemonion*, which would mean "floret", and he cites several examples of related Greek words (but not that one) which describe chemical or biological efflorescence.

The early uses of *antimonium* include the translations, in 1050–1100, by Constantine the African of Arabic medical treatises.

Several authorities believe that *antimonium* is a scribal corruption of some Arabic form; Meyerhof derives it from *ithmid*; other possibilities include *Athimar*, the Arabic name of the *metal*, and a hypothetical *as-stimmi*, derived from or parallel to the Greek.

Characteristics

Figure 66: A vial containing a black allotrope of antimony

Figure 67: Native antimony with oxidation products

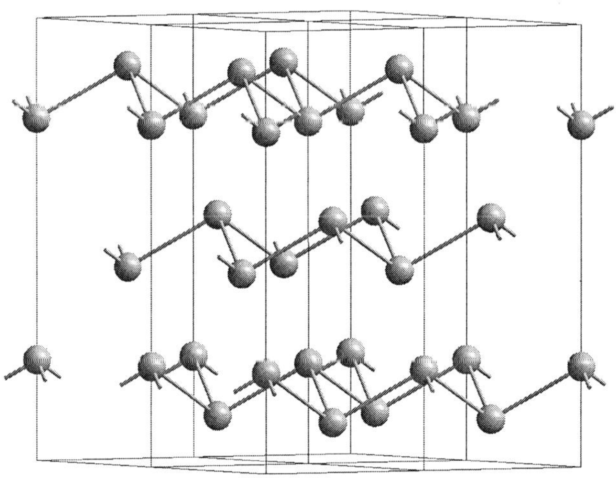

Figure 68: Crystal structure common to Sb, as Sb and grey As

Antimony is in the nitrogen group (group 15) and has an electronegativity of 2.05. As expected by periodic trends, it is more electronegative than tin or bismuth, and less electronegative than tellurium or arsenic.

Antimony is stable in air at room temperature but reacts with oxygen if heated to form antimony trioxide, Sb_2O_3.

Antimony is a silvery, lustrous gray metal that has a Mohs scale hardness of 3. Therefore, antimony by itself is not used to make hard objects: coins made of antimony were issued in China's Guizhou province in 1931, but because of their rapid wear their minting was discontinued. Antimony is resistant to attack by acids.

Four allotropes of antimony are known: a stable metallic form, and three metastable forms: explosive, black and yellow. Metallic antimony is a brittle, silver-white shiny metal. When molten antimony is slowly cooled, metallic antimony crystallises in an hexagonal cell, isomorphic with that of the grey allotrope of arsenic.

A rare explosive form of antimony can be formed from the electrolysis of antimony(III) trichloride. When scratched with a sharp implement, an exothermic reaction occurs and white fumes given off as metallic antimony is formed; alternatively, when rubbed with a pestle in a mortar, a strong detonation occurs.

Black antimony is formed upon rapid cooling of gaseous metallic antimony. It has the same crystal structure as red phosphorus and black arsenic, it oxidises in air and may ignite spontaneously. At 100 °C, it gradually transforms into the stable form. The yellow allotrope of antimony is the most unstable. It has only been generated by oxidation of stibine (SbH_3) at "90 °C. Above this temperature and in ambient light, this meta stable allotrope transforms into the more stable black allotrope.

Metallic antimony adopts a double-layered structure (space group R3m No. 166) consisting of many interlocked ruffled six-membered rings. Nearest and next-nearest neighbours form a distorted octahedral complex, with the three atoms in the same double-layer being slightly closer than the three atoms in the next. This relatively close packing leads to a high density of 6.697 g/cm^3 whereas the low hardness and brittleness of antimony originate from the weak bonding among the layers.

Isotopes

Antimony exists as two stable isotopes, ^{121}Sb with a natural abundance of 57.36% and ^{123}Sb with a natural abundance of 42.64%. It also has 35 radioisotopes, of which the longest-lived is ^{125}Sb with a half-life of 2.75 years. In addition, 29 metastable states have been characterised.

Occurrence

The abundance of antimony in the Earth's crust is estimated at 0.2 to 0.5 parts per million, comparable to thallium at 0.5 parts per million and silver at 0.07 ppm. Even though this element is not abundant, it is found in over 100 mineral species. Antimony is sometimes found native, but more frequently it is found in the sulfide stibnite (Sb_2S_3) which is the predominant ore mineral. Commercial forms of antimony are generally ingots, broken pieces, granules, and cast cake. Other forms are powder, shot, and single crystals.

In 2005, the People's Republic of China was the top producer of antimony with about 84% world share followed at a distance by South Africa, Bolivia and Tajikistan, reports the British Geological Survey. The mine with the largest deposits in China is Xikuangshan Mine in Hunan province with a estimated deposit of 2.1 million metric tons. In October 2011 a deposit of antimony was found in a shallow seabed about 50 km off Amami-Oshima Island in Kagoshima Prefecture. The discovery was the first time that antimony had been found at such shallow depths(480 metres), with this type of mineral deposit only ever having been found in depths in excess of 1000 metres.

Production

The extraction of antimony from ores depends on the quality of the ore, which is usually a sulfide. The sulfide is converted to an oxide and advantage is often taken of the volatility of antimony(III) oxide, which is recovered from roasting. This material is often used directly for the main applications, impurities being arsenic and sulfide. Antimony can be isolated from its ore by a reduction with scrap iron:

$$Sb_2S_3 + 3\ Fe \rightarrow 2\ Sb + 3\ FeS$$

Isolating antimony from its oxide is performed by a carbothermal reduction:

$$2\ Sb_2O_3 + 3\ C \rightarrow 4\ Sb + 3\ CO_2$$

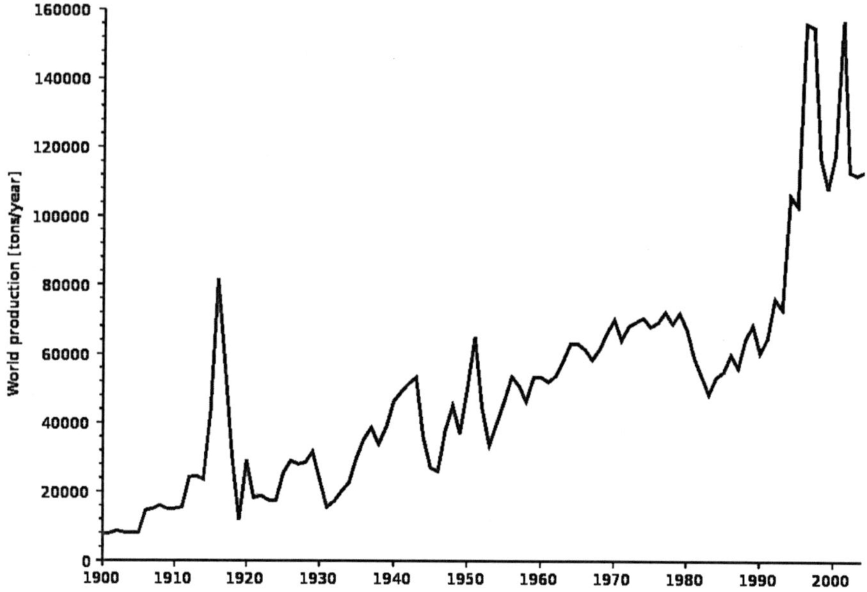

Figure 69: World production trend of antimony

Table 1: Antimony production in 2010

Country	*Tonnes*	*% of total*
People's Republic of China	120,000	88.9
South Africa	3,000	2.2
Bolivia	3,000	2.2
Russia	3,000	2.2
Tajikistan	2,000	1.5
Top 5	131,000	97.0
Total world	135,000	100.0

Compounds

Antimony compounds are often classified into those of Sb(III) and Sb(V). Relative to its neighbouring element As, the 5+ oxidaton state is more stable.

Oxides and Hydroxides

Antimony trioxide (Sb_4O_6) is formed when antimony is burnt in air. In the gas phase, this compound exists as Sb_4O_6, but it polymerises upon condensing. Antimony pentoxide, (Sb_4O_{10}) can only be formed by oxidation by concentrated nitric acid. Antimony also forms a mixed-valence oxide, antimony tetroxide (Sb_2O_4), which features both Sb(III) and Sb(V). Unlike phosphorus and arsenic, these various oxides are amphoteric and do not form well-defined oxoacids and react with acids to form antimony salts.

Antimonous acid $Sb(OH)_3$ is unknown but the conjugate base sodium antimonite ($[Na_3SbO_3]_4$) forms upon fusing sodium oxide and Sb_4O_6. Transition metal antimonites are also known. Antimonic acid exists only as the hydrate $HSb(OH)_6$, forming salts containing the antimonate anion $Sb(OH)^-$

6. Dehydrating metal salts containing this anion yields mixed oxides.

Many antimony ores are sulfides, including stibnite (Sb_2S_3), pyrargyrite (Ag_3SbS_3), zinkenite, jamesonite, and boulangerite. Antimony pentasulfide is non-stoichiometric and features antimony in the +3 oxidation state and S-S bonds. Several thioantimonides are known such as $[Sb_6S_{10}]^{2-}$ and $[Sb_8S_{13}]^{2-}$.

Halides

Antimony forms two series of halides, SbX_3 and SbX_5. The trihalides SbF_3, $SbCl_3$, $SbBr_3$, and SbI_3 are all molecular compounds having trigonal pyramidal molecular geometry. The trifluoride SbF_3 is prepared by the reaction of Sb_2O_3 with HF:

$$Sb_2O_3 + 6\ HF \rightarrow 2\ SbF_3 + 3\ H_2O$$

It is Lewis acidic and readily accepts fluoride ions to form the complex anions SbF^-4 and $SbF2^-$

5. Molten SbF_3 is a weak electrical conductor. The trichloride $SbCl_3$ is prepared by dissolving Sb_2S_3 in hydrochloric acid:

$$Sb_2S_3 + 6\ HCl \rightarrow 2\ SbCl_3 + 3\ H_2S$$

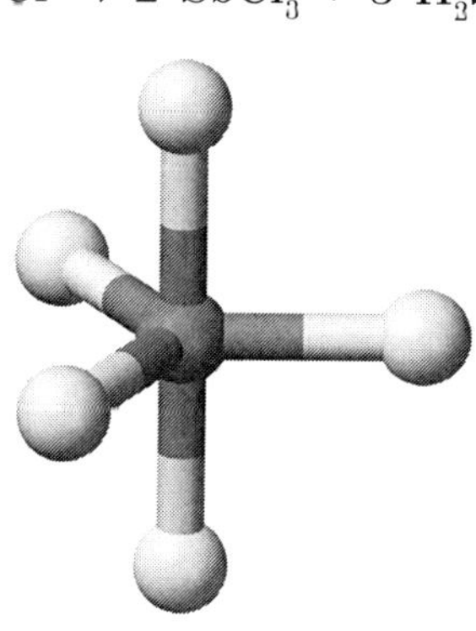

Figure 70: Structure of gaseous SbF_5

The pentahalides SbF_5 and $SbCl_5$ have trigonal bipyramidal molecular geometry in the gas phase, but in the liquid phase, SbF_5 is polymeric, whereas $SbCl_5$ is monomeric. SbF_5 is a powerful Lewis acid used to make the superacid fluoroantimonic acid ("$HSbF_6^-$).

Oxyhalides are more common for antimony than arsenic and phosphorus. Antimony trioxide dissolves in concentrated acid to form antimony oxo- (antimonyl) compounds such as SbOCl and $(SbO)_2SO_4$.

[.] Antimonides, hydrides, and organoantimony compounds

Compounds in this class generally are described as derivatives of Sb^{3-}. Antimony forms *antimonides* with metals, such as indium antimonide (InSb), and silver antimonide (Ag_3Sb). The alkali metal and zinc antimonides, e.g. Na_3Sb and Zn_3Sb_2, are more reactive. Treating these antimonides with acid produces the unstable gas stibine, SbH_3:

$$Sb^{3-} + 3\ H^+ \rightarrow SbH_3$$

Stibine can also be produced by treating Sb^{3+} salts with hydride reagents such as sodium borohydride. Stibine decomposes spontaneously at room temperature. Because stibine is thermodynamically unstable (positive heat of formation), antimony does not react with hydrogen directly.

Organoantimony compounds are typically prepared by alkylation of antimony halides with Grignard reagents. A large variety of compounds are known with both Sb(III) and Sb(V) centres including mixed chloro-organic derivatives, anions, and cations. Examples include $Sb(C_6H_5)_3$ (triphenylstibine), $Sb_2(C_6H_5)_4$ (with an Sb-Sb bond), and cyclic $[Sb(C_6H_5)]_n$. Pentacoordinated organoantimony compounds are common, examples being $Sb(C_6H_5)_5$ and several related halides.

Applications

Flame Retardants

The main use of antimony is in the form of antimony trioxide is used in the making of flame-proofing compounds. Markets for these flame-retardant applications include children's clothing, toys, aircraft and automobile seat covers. It is also used in the fibreglass composites industry as an additive to polyester resins for such items as light aircraft engine covers. The resin will burn while a flame is held to it but will extinguish itself as soon as the flame is removed. Fireproofing consumes about half of the annual production of antimony.

Alloys

Antimony forms a highly useful alloy with lead, increasing its hardness and mechanical strength. The Sb–Pb alloy is used in lead–acid batteries. It is used in antifriction alloys, such as Babbitt metal. It is used as an alloy in bullets and lead shot, cable sheathing, type metal (e.g. for linotype printing machines), solder – some "lead-free" solders contain 5% Sb, in pewter, and in hardening alloys with low tin content in the manufacturing of organ pipes.

Other Applications

The second main application is as a catalyst for the production of the polymer polyethyleneterephthalate. It is an additive in some glasses. In the latter application, antimony oxides serve as fining agents, aiding in the removal of microscopic bubbles. This application is mainly used for TV screens.

Niche Applications

Microelectronics

In tiny amounts, antimony is increasingly being used in the semiconductor industry as a dopant for ultra-high conductivity n-type silicon wafers in the production of diodes, infrared detectors, and Hall-effect devices.

In the 1950s, tiny beads of a lead-antimony alloy were used to dope the emitters and collectors of NPN alloy junction transistors with antimony.

Medical

Few biological or medical applications exist for antimony. Treatments principally containing antimony are known as antimonials and are used as emetics.

Antimony compounds are used as antiprotozoan drugs. Antimony potassium tartrate, or tartar emetic, was once used as an anti-schistosomal drug, subsequently replaced by praziquantel.

Antimony and its compounds are used in several veterinary preparations like anthiomaline or lithium antimony thiomalate, which is used as a skin conditioner in ruminants.

Antimony has a nourishing or conditioning effect on keratinized tissues, at least in animals. Antimony-based drugs, such as meglumine antimoniate, are also considered the drugs of choice for treatment of leishmaniasis in domestic animals. Unfortunately, as well as having low therapeutic indices, the drugs are poor at penetrating the bone marrow, where some of the Leishmania *amastigotes reside, and so cure of the disease – especially the visceral form – is very difficult.*

Elemental antimony as an antimony pill was once used as a medicine. It could be reused by others after ingestion.

Other Uses

In the heads of some safety matches and in nuclear reactors together with beryllium as startup neutron sources.

Antimony sulfides have been shown to help stabilise the friction coefficient in automotive brake pad materials. Antimony also is used in the making of bullets and bullet tracers. This element is also used in cosmetics and event paint and glass art crafts.

Precautions

Antimony and many of its compounds are toxic, and the effects of antimony poisoning are similar to arsenic poisoning. Inhalation of antimony dust is harmful and in certain cases may be fatal; in small doses, antimony causes headaches, dizziness, and depression. Larger doses such as prolonged skin contact may cause dermatitis; otherwise it can damage the kidneys and the liver, causing violent and frequent vomiting, and will lead to death in a few days.

Antimony is incompatible with strong oxidising agents, strong acids, halogen acids, chlorine, or fluorine. It should be kept away from heat.

Antimony leaches from polyethylene terephthalate (PET) bottles into liquids. While levels observed for bottled water are below drinking water guidelines, fruit juice concentrates (for which no guidelines are established) produced in the UK were found to contain up to 44.7 µg/L of antimony, well above the EU limits for tap water of 5 µg/L. The guidelines are:

- World Health Organisation: 20 µg/L
- Japan: 15 µg/L
- United States Environmental Protection Agency, Health Canada and the Ontario Ministry of Environment: 6 µg/L
- German Federal Ministry of Environment: 5 µg/L

Antitaenite

Antitaenite is a meteoritic metal alloy mineral composed of iron and nickel, 20-40% Ni (and traces of other elements) that has a face centred cubic crystal structure. Its existence as a new mineral species occurring in both iron meteorites and in chondrites was first recognised in 1995. There are three other known Fe-Ni metreotic minerals: kamacite, taenite, and tetrataenite. The pair of minerals antitaenite and taenite constitute the first example in nature of two minerals that have the same crystal structure (face centred cubic) and can have the same chemical composition (same proportions of Fe and Ni) - they differ in their electronic structures: taenite is a high magnetic moment alloy whereas antitaenite is a low magnetic moment alloy. This unique difference in electronic structure was first established in 1999 and arises from a high-magnetic-moment to low-magnetic-moment transition occurring in the Fe-Ni bi-metallic alloy series. The same electronic structure transition is believed to be a causal factor in Invar behaviour.

Antlerite

Figure 71: A vein of acicular, green antlerite crystals from Chuquicamata, Chile (dimensions: 3.5 x 3.4 x 1.8 cm)

Antlerite is a greenish hydrous copper sulfate mineral, with the formula $Cu_3(SO_4)(OH)_4$. It occurs in tabular, acicular, or fibrous crystals with a vitreous luster. Originally believed to be a rare mineral, antlerite was found to be the primary ore of the oxidised zones in several copper mines across the world, including the Chuquicamata mine in Chile, and the Antler mine in Arizona, USA from which it takes its name. It is chemically and optically similar in many respects to other copper minerals such as malachite and brochantite, though it can be distinguished from the former by a lack of effervescence in hydrochloric acid.

Apatite

Figure 72: Apatite

Apatite is a group of phosphate minerals, usually referring to hydroxylapatite, fluorapatite, chlorapatite and bromapatite, named for high concentrations of OH^-, F^-, Cl^- or Br^- ions, respectively, in the crystal. The formula of the admixture of the four most common endmembers is written as $Ca_{10}(PO_4)_6(OH,F,Cl,Br)_2$, and the crystal unit cell formulae of the individual minerals are written as $Ca_{10}(PO_4)_6(OH)_2$, $Ca_{10}(PO_4)_6(F)_2$, $Ca_{10}(PO_4)_6(Cl)_2$ and $Ca_{10}(PO_4)_6(Br)_2$.

Apatite is one of a few minerals that are produced and used by biological micro-environmental systems. Apatite is the defining mineral for 5 on the Mohs scale. Hydroxyapatite, also known as hydroxylapatite, is the major component of tooth enamel and bone mineral. A relatively rare form of apatite in which most of the OH groups are absent and containing many carbonate and acid phosphate substitutions is a large component of bone material.

Fluorapatite (or fluoroapatite) is more resistant to acid attack than is hydroxyapatite. For this reason, toothpaste typically contains a source of fluoride anions (e.g. sodium fluoride, sodium monofluorophosphate). Similarly, fluoridated water allows exchange in the teeth of fluoride ions for hydroxyl groups in apatite. Too much fluoride results in dental fluorosis and/or skeletal fluorosis.

Fission tracks in apatite are commonly used to determine the thermal history of orogenic (mountain) belts and of sediments in sedimentary basins. (U-Th)/He dating of apatite is also well established for use in determining thermal histories and other, less typical applications such as paleo-wildfire dating.

Phosphorite is a phosphate-rich sedimentary rock, that contains between 18% and 40% P_2O_5. The apatite in phosphorite is present as cryptocrystalline masses referred to as

Uses

The primary use of apatite is in the manufacture of fertilizer – it is a source of phosphorus. It is occasionally used as a gemstone.

During digestion of apatite with sulfuric acid to make phosphoric acid, hydrogen fluoride is produced as a byproduct from any fluorapatite content. This byproduct is a minor industrial source of hydrofluoric acid. Fluoro-chloro apatite forms the basis of the now obsolete Halophosphor fluorescent tube phosphor system. Dopant elements of manganese and antimony, at less than one mole-percent, in place of the calcium and phosphorus impart the fluorescence, and adjustment of the fluorine to chlorine ratio adjusts the shade of white produced. Now almost entirely replaced by the Tri-Phosphor system.

In the United States, apatite derived fertilizers are used to supplement the nutrition of many agricultural crops by providing a valuable source of phosphate.

Gemology

Apatite is infrequently used as a gemstone. Transparent stones of clean colour have been faceted, and chatoyant specimens have been cabochon cut. Chatoyant stones are known as *cat's-eye apatite*, transparent green stones are known as *asparagus stone*, and blue stones have been called *moroxite*. Crystals of rutile may have grown in the crystal of apatite so when in the right light, the cut stone displays a cat's eye effect. Major sources for gem apatite are Brazil, Burma, and Mexico. Other sources include Canada, Czechoslovakia, Germany, India, Madagascar, Mozambique, Norway, South Africa, Spain, Sri Lanka, and the United States.

Use as an Ore Mineral

Apatite is occasionally found to contain significant amounts of rare earth elements and can be used as an ore for those metals. This is preferable to traditional rare earth ores, as apatite is non-radioactive and does not pose an environmental hazard in mine tailings. Except some apatite in Florida used to produce phosphate for U.S. tobacco crops contains uranium, radium, lead 210 and polonium 210 and radon.

Apatite is an ore mineral at the Hoidas Lake rare earth project.

Thermodynamics

The standard (p = 0.1 MPa) molar enthalpies of formation in the crystalline state of hydroxyapatite, chlorapatite and a preliminary value for bromapatite, at T = 298.15 K, have already been determined by reaction-solution calorimetry. Speculations on the existence of a possible fifth member of the calcium apatites family, iodoapatite, have been drawn from energetic considerations.

Lunar Science

Moon rocks collected by astronauts during the Apollo program contain traces of apatite. Re-analysis of these samples in 2010 revealed water trapped in the mineral as hydroxyl, leading to estimates of water on the lunar surface at a rate of at least 64 parts per billion – 100 times greater than previous estimates – and as high as 5 parts per million. If the minimum amount of mineral-locked water was hypothetically converted to liquid, it would cover the Moon's surface in roughly one metre of water.

Apophyllite

Figure 73: Apophyllite

The name apophyllite refers to a specific group of phyllosilicates, a class of minerals that also includes the micas. Originally, the group name referred to a specific mineral, but was redefined in 1978 to stand for a class of minerals of similar chemical makeup that comprise a solid solution series, and includes the members apophyllite-(KF), apophyllite-(KOH), and apophyllite-(NaF). These minerals are typically found as secondary minerals in holes in basalt or other igneous rocks. They can also be called "fisheye stone". A recent change in the nomenclature system used for this group was approved by the International Mineralogical Association, removing the prefixes from the species names and using suffixes to designate the species.

Though relatively unfamiliar to the general public, apophyllites are fairly prevalent around the world, with specimens coming from some of the world's most well-known mineral localities. These localities include: Poona, India; the Harz Mountains of Germany, Mont Saint-Hilaire in Canada, and Kongsberg, Norway, with other locations in Scotland, Ireland, Brazil, Japan, and throughout the United States.

Uses

Apophyllites are popular as collector's minerals. This popularity is due to a combination of factors, including their abundance, colour variety, and well-defined crystals. Naturally forming pyramidal structures, they refract light in obvious rainbows, and can form "natural pyramids" when subjected to rock tumbling.

Species of Apophyllite

- Apophyllite-(KF) (formerly fluorapophyllite), $(K, Na)Ca_4Si_8O_{20}(F, OH) \cdot 8H_2O$ - white, colourless, yellow, green, violet
- Apophyllite-(KOH) (formerly hydroxyapophyllite), $KCa_4Si_8O_{20}(OH, F) \cdot 8H_2O$ - white, colourless
- Apophyllite-(NaF) (formerly natroapophyllite), $(K, Na)Ca_4Si_8O_{20}F \cdot 8H_2O$ - brown, yellow, colourless

Aragonite

Aragonite is a carbonate mineral, one of the two common, naturally occurring, crystal forms of calcium carbonate, $CaCO_3$ (the other form being the mineral calcite). It is formed by biological and physical processes, including precipitation from marine and freshwater environments.

Figure 74: Aragonite from Salsignes Mine, Aude department, France Size: 30x30x20 cm

Aragonite's crystal lattice differs from that of calcite, resulting in a different crystal shape, an orthorhombic system with acicular crystals. Repeated twinning results in pseudo-hexagonal forms. Aragonite may be columnar or fibrous, occasionally in branching stalactitic forms called *flos-ferri* ("flowers of iron") from their association with the ores at the Carinthian iron mines.

Occurrence

The type location for aragonite is Molina de Aragón (Guadalajara, Spain), 25 km from Aragon for which it was named in 1797. An aragonite cave, the Ochtinská Aragonite Cave, is situated in Slovakia. In the USA, aragonite in the form of stalactites and "cave flowers" (anthodite) is known from Carlsbad Caverns and other caves. Massive deposits of oolitic aragonite sand are found on the seabed in the Bahamas.

Aragonite forms naturally in almost all mollusk shells, and as the calcareous endoskeleton of warm- and cold-water corals (Scleractinia). Because the mineral deposition in mollusk shells is strongly biologically controlled, some crystal forms are distinctively different from those of inorganic aragonite. In some mollusks, the entire shell is aragonite; in others, aragonite forms only discrete parts of a bimineralic shell (aragonite plus calcite). Aragonite also forms in the ocean and in caves as inorganic precipitates called marine cements and speleothems, respectively. The nacreous layer of the aragonite fossil shells of some extinct ammonites forms an iridescent material called ammolite. Ammolite is primarily aragonite with impurities that make it iridescent and valuable as a gemstone.

Aragonite is metastable and is thus commonly replaced by calcite in fossils. Aragonite older than the Carboniferous is essentially unknown.

Physical Properties

Aragonite is thermodynamically unstable at standard temperature and pressure, and tends to alter to calcite on scales of 10^7 to 10^8 years. The mineral vaterite, also known as ì-$CaCO_3$, is another phase of calcium carbonate that is metastable at ambient conditions typical of Earth's surface, and decomposes even more readily than aragonite.

Figure 75: Aragonite crystals from Cuenca, Castile-La Mancha, Spain

Figure 76: Cluster of twinned aragonite from Morocco

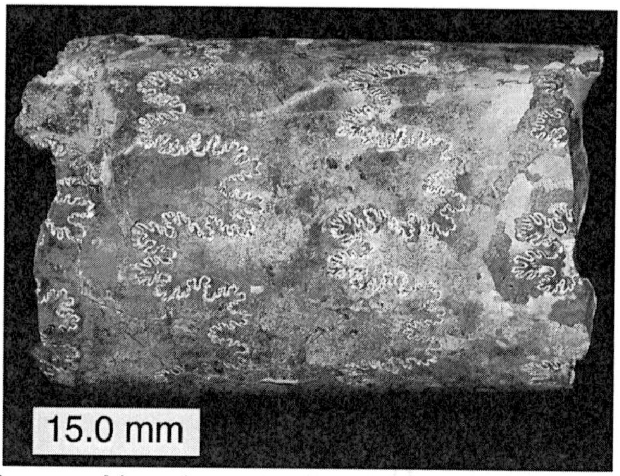

Figure 77: Remnant biogenic aragonite (thin, rainbow-coloured shell) on the ammonite *Baculites* (Pierre Shale, Late Cretaceous, South Dakota).

Arctite

Arctite ($Na_2Ca_4(PO_4)_3F$) is a colourless mineral found in the Kola Peninsula northern Russia. It's Mohs hardness is 5 and has a density of 3.11-3.13. Arctite is transparent with a vitreous lustre. Arctite has a perfect cleavage and a trigonal crystal system.

Common associates of arctite include aegirine, natisite, lomonosovite, umbite and thenardite.

Arfvedsonite

Figure 78: Arfvedsonite from New Hampshire, USA

Arfvedsonite is a sodium amphibole mineral with composition: $[Na][Na_2][(Fe^{2+})_4Fe^{3+}][(OH)_2 | Si_8O_{22}]$. It crystallises in the monoclinic prismatic crystal system and typically occurs as greenish black to bluish grey fibrous to radiating or stellate prisms.

It is a rather rare mineral occurring in nepheline syenite intrusions and agpaitic (peralkaline) pegmatites and granites as the Golden Horn batholith in Okanogan County, Washington (type locality for zektzerite).

Occurrences include Mont Saint-Hilaire, Quebec, Canada; the Ilímaussaq complex in Southern Greenland; and in pegmatites of the Kola Peninsula, Russia. Its mineral association includes nepheline, albite, aegirine, riebeckite, katophorite and quartz.

Arfvedsonite was discovered in 1823 and named for the Swedish chemist Johan August Arfwedson (1792–1841).

Argutite

Argutite (GeO_2) is a rare germanium oxide mineral. It is a member of the rutile group.

It was first described for an occurrence in the Argut deposit, central Pyrenees, Haute-Garonne, France in 1983. The type locality is within a zinc ore deposit within lower Paleozoic sedimentary rocks that have undergone metamorphism. Associated minerals include sphalerite, cassiterite, siderite and briartite.

Armalcolite

Armalcolite is a titanium-rich mineral with the chemical formula $(Mg,Fe^{2+})Ti_2O_5$. It was first found at Tranquility Base on the Moon in 1969 and named for Armstrong, Aldrin and Collins, the three Apollo 11 astronauts.

Together with tranquillityite and pyroxferroite, it is one of the three minerals which were discovered on the Moon. Armalcolite was then identified in various locations on Earth and synthesized in the laboratory.

The synthesis requires low pressures, high temperatures and rapid quenching from about 1000 °C to the ambient temperature. Armalcolite breaks down to the mixture of magnesium-rich ilmenite and rutile at temperatures below 1000 °C, but the conversion slows down with cooling. Because of this quenching requirement, armalcolite is relatively rare and is usually found in association with ilmenite and rutile, among other minerals.

Arsenic

Figure 79: Metallic grey

Arsenic is a chemical element with the symbol As, atomic number 33 and relative atomic mass 74.92.

Arsenic occurs in many minerals, usually in conjunction with sulfur and metals, and also as a pure elemental crystal. It was first documented by Albertus Magnus in 1250.

Arsenic is a metalloid. It can exist in various allotropes, although only the grey form has important use in industry.

The main use of metallic arsenic is for strengthening alloys of copper and especially lead (for example, in car batteries).

Arsenic is a common n-type dopant in semiconductor electronic devices, and the optoelectronic compound gallium arsenide is the most common semiconductor in use after doped silicon.

A few species of bacteria are able to use arsenic compounds as respiratory metabolites, and are arsenic-tolerant.

Arsenic is notoriously poisonous to multicellular life because of the interaction of arsenic ions with protein thiols. Arsenic and its compounds, especially the trioxide, are used in the production of pesticides (treated wood products), herbicides, and insecticides.

These applications are declining, however, as many of these compounds are being phased out. Arsenic poisoning from naturally occurring arsenic compounds in drinking water remains a problem in many parts of the world.

Characteristics

The three most common allotropes are *metallic grey*, *yellow* and *black arsenic*, with grey being the most common. *Grey arsenic* (α-As, space group R3m No. 166) adopts a double-layered structure consisting of many interlocked ruffled six-membered rings.

Because of weak bonding between the layers, grey arsenic is brittle and has a relatively low Mohs hardness of 3.5. Nearest and next-nearest neighbours form a distorted octahedral complex, with the three atoms in the same double-layer being slightly closer than the three atoms in the next.

This relatively close packing leads to a high density of 5.73 g/cm^3. Grey arsenic is a semimetal, but becomes a semiconductor with a bandgap of 1.2–1.4 eV if amorphized. *Yellow arsenic* is soft and waxy, and somewhat similar to tetraphosphorus (P_4).

Both have four atoms arranged in a tetrahedral structure in which each atom is bound to each of the other three atoms by a single bond. This unstable allotrope, being molecular, is the most volatile, least dense and most toxic. Solid yellow arsenic is produced by rapid cooling of arsenic vapour, As_4. It is rapidly transformed into the grey arsenic by light. The yellow form has a density of 1.97 g/cm^3. *Black arsenic* is similar in structure to red phosphorus.

Isotopes

Naturally occurring arsenic is composed of one stable isotope, ^{75}As. As of 2003, at least 33 radioisotopes have also been synthesized, ranging in atomic mass from 60 to 92. The most stable of these is ^{73}As with a half-life of 80.3 days. Isotopes that are lighter than the stable ^{75}As tend to decay by β^+ decay, and those that are heavier tend to decay by β^- decay, with some exceptions.

At least 10 nuclear isomers have been described, ranging in atomic mass from 66 to 84. The most stable of arsenic's isomers is ^{68m}As with a half-life of 111 seconds.

Chemistry

When heated in air, arsenic oxidises to arsenic trioxide; the fumes from this reaction have an odour resembling garlic. This odour can be detected on striking arsenide minerals such as arsenopyrite with a hammer. Arsenic (and some arsenic compounds) sublimes upon heating at atmospheric pressure, converting directly to a gaseous form without an intervening liquid state at 887 K (614 °C). The triple point is 3.63 MPa and 1,090 K (820 °C). Arsenic makes arsenic acid with concentrated nitric acid, arsenious acid with dilute nitric acid, and arsenic trioxide with concentrated sulfuric acid.

Compounds

Arsenic compounds resemble in some respects those of phosphorus, which occupies the same group (column) of the periodic table. Arsenic is less commonly observed in the pentavalent state, however. The most common oxidation states for arsenic are: "3 in the arsenides, such as alloy-like intermetallic compounds; and +3 in the arsenites, arsenates(III), and most organoarsenic compounds. Arsenic also bonds readily to itself as seen in the square As3⁻4 ions in the mineral skutterudite. In the +3 oxidation state, arsenic is typically pyramidal, owing to the influence of the lone pair of electrons.

Inorganic

Arsenic forms colourless, odourless, crystalline oxides As_2O_3 ("white arsenic") and As_2O_5, which are hygroscopic and readily soluble in water to form acidic solutions. Arsenic(V) acid is a weak acid. Its salts are called arsenates, e.g., Paris Green (copper(II) acetoarsenite), calcium arsenate, and lead hydrogen arsenate. The latter three have been used as agricultural insecticides and poisons. The protonation steps between the arsenate and arsenic acid are similar to those between phosphate and phosphoric acid. However, arsenite and arsenous acid contain arsenic bonded to three oxygen and not hydrogen atoms, in contrast to phosphite and phosphonic acid, which contains a non-acidic P-H bond. Arsenous acid is genuinely tribasic, whereas phosphonic acid is not.

A broad variety of sulfur compounds of arsenic are known. Orpiment (As_2S_3) and realgar (As_4S_4) are somewhat abundant and were formerly used as painting pigments. Other sulfides include As_4S_3 and As_4S_{10}. Arsenic has a formal oxidation state of +2 in As_4S_4, which features As-As bonds so that the total covalency of As is still in fact three.

The trifluoride, trichloride, tribromide, and triiodide of arsenic(III) are well known, whereas only AsF_5 is the main pentahalide. Arsenic pentafluoride is stable at room temperature, whereas the pentachloride is stable only below "50 °C.

Organoarsenic Compounds

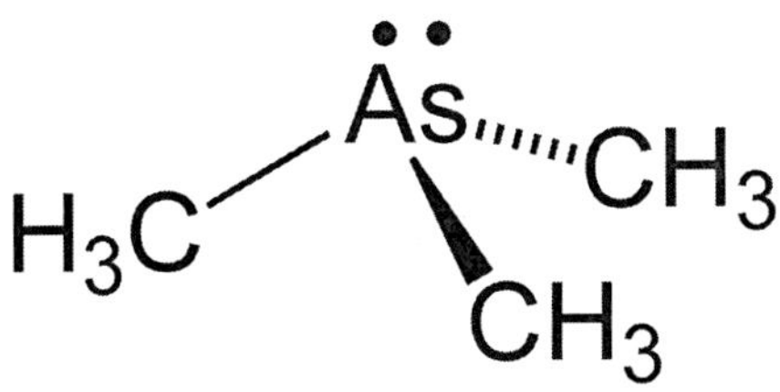

Figure 80: Trimethylarsine

A large variety of organoarsenic compounds are known. Several were developed as chemical warfare agents during World War I, including vesicants such as lewisite and vomiting agents such as adamsite. Cacodylic acid, which is of historic and practical interest, arises from the methylation of arsenic trioxide, a reaction that has no analogy in phosphorus chemistry.

Alloys

Arsenic is used as the group 5 element in the III-V semiconductors gallium arsenide, indium arsenide, and aluminium arsenide. The valence electron count of GaAs is the same as a pair of Si atoms, but the band structure is completely different, which results distinct bulk properties. Other arsenic alloys include the II-IV semiconductor cadmium arsenide.

Occurrence and Production

Figure 81: A large sample of native arsenic.

Minerals with the formula MAsS and MAs_2 (M = Fe, Ni, Co) are the dominant commercial sources of arsenic, together with realgar (an arsenic sulfide mineral) and native arsenic. An illustrative mineral is arsenopyrite (FeAsS), which is structurally related to iron pyrite. Many minor As-containing minerals are known. Arsenic also occurs in various organic forms in the environment. Inorganic arsenic and its compounds, upon entering the food chain, are progressively metabolised to a less toxic form of arsenic through a process of methylation.

Other naturally occurring pathways of exposure include volcanic ash, weathering of arsenic-containing minerals and ores, and dissolved in groundwater. It is also found in food, water, soil, and air. The most common pathway of exposure for humans is ingestion, and the predominant source of arsenic in our diet is through seafood. An additional route of exposure is through inhalation.

In 2005, China was the top producer of white arsenic with almost 50% world share, followed by Chile, Peru, and Morocco, according to the British Geological Survey and the United States Geological Survey. Most operations in the US and Europe have closed for environmental reasons. The arsenic is recovered mainly as a side product from the purification of copper. Arsenic is part of the smelter dust from copper, gold, and lead smelters.

On roasting in air of arsenopyrite, arsenic sublimes as arsenic(III) oxide leaving iron oxides, while roasting without air results in the production of metallic arsenic. Further purification from sulfur and other chalcogens is achieved by sublimation in vacuum or in a hydrogen atmosphere or by distillation from molten lead-arsenic mixture.

History

Figure 82: Realgar

Figure 83: Alchemical symbol for arsenic

It is also related to the similar Greek word *arsenikos*, meaning "masculine" or "potent". The word was adopted in Latin *arsenicum* and Old French *arsenic,* from which the English word *arsenic* is derived. Arsenic sulfides (orpiment, realgar) and oxides have been known and used since ancient times. Zosimos (circa 300 AD) describes roasting *sandarach* (realgar) to obtain *cloud of arsenic* (arsenious oxide), which he then reduces to metallic arsenic. As the symptoms of arsenic poisoning were somewhat ill-defined, it was frequently used for murder until the advent of the Marsh test, a sensitive chemical test for its presence. (Another less sensitive but more general test is the Reinsch test.) Owing to its use by the ruling class to murder one another and its potency and discreetness, arsenic has been called the *Poison of Kings* and the *King of Poisons.*

During the Bronze Age, arsenic was often included in bronze, which made the alloy harder (so-called "arsenical bronze"). Albertus Magnus (Albert the Great, 1193–1280) is believed to have been the first to isolate the element in 1250 by heating soap together with arsenic trisulfide. In 1649, Johann Schröder published two ways of preparing arsenic.

Cadet's fuming liquid (impure cacodyl), often claimed as the first synthetic organometallic compound, was synthesized in 1760 by Louis Claude Cadet de Gassicourt by the reaction of potassium acetate with arsenic trioxide.

In the Victorian era, "arsenic" ("white arsenic" trioxide) was mixed with vinegar and chalk and eaten by women to improve the complexion of their faces, making their skin paler to show they did not work in the fields. Arsenic was also rubbed into the faces and arms of women to "improve their complexion". The accidental use of arsenic in the adulteration of foodstuffs led to the Bradford sweet poisoning in 1858, which resulted in approximately 20 deaths.

Applications

Agricultural:

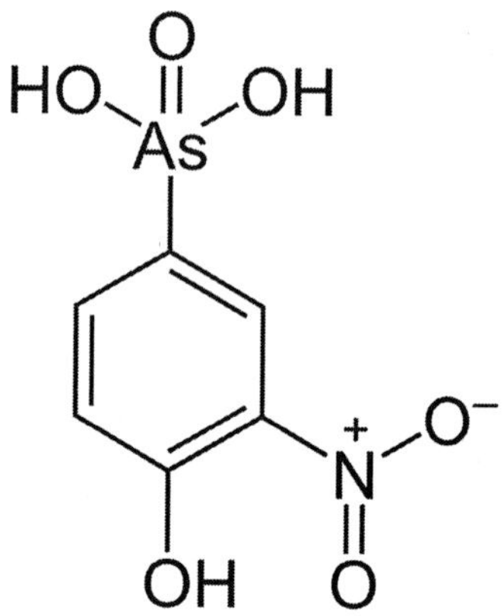

Figure 84: Roxarsone is a controversial arsenic compound used as a nutritional supplement for chickens.

The toxicity of arsenic to insects, bacteria and fungi led to its use as a wood preservative. In the 1950s a process of treating wood with chromated copper arsenate (also known as CCA or Tanalith) was invented, and for decades this treatment was the most extensive industrial use of arsenic. An increased appreciation of the toxicity of arsenic resulted in a ban for the use of CCA in consumer products; the European Union and United States initiated this process in 2004. CCA remains in heavy use in other countries however, e.g. Malaysian rubber plantations.

Arsenic was also used in various agricultural insecticides, termination and poisons. For example, lead hydrogen arsenate was a common insecticide on fruit trees, but contact with the compound sometimes resulted in brain damage among those working the sprayers. In the second half of the 20th century, monosodium methyl arsenate (MSMA) and disodium methyl arsenate (DSMA) – less toxic organic forms of arsenic – have replaced lead arsenate in agriculture.

Arsenic is still added to animal food, in particular in the U.S. as a method of disease prevention and growth stimulation. One example is roxarsone, which is used as a broiler starter by about 70% of the broiler growers since 1995. The Poison-Free Poultry Act of 2009 proposes to ban the use of roxarsone in industrial swine and poultry production.

Medical Use

During the 18th, 19th, and 20th centuries, a number of arsenic compounds have been used as medicines, including arsphenamine (by Paul Ehrlich) and arsenic trioxide (by Thomas Fowler). Arsphenamine as well as neosalvarsan was indicated for syphilis and trypanosomiasis, but has been superseded by modern antibiotics. Arsenic trioxide has been used in a variety of ways over the past 500 years, but most commonly in the treatment of cancer.

The US Food and Drug Administration in 2000 approved this compound for the treatment of patients with acute promyelocytic leukemia that is resistant to ATRA. It was also used as Fowler's solution in psoriasis.

Recently new research has been done in locating tumours using arsenic-74 (a positron emitter). The advantages of using this isotope instead of the previously used iodine-124 is that the signal in the PET scan is clearer as the body tends to transport iodine to the thyroid gland producing a lot of noise.

In subtoxic doses, soluble arsenic compounds act as stimulants, and were once popular in small doses as medicine by people in the mid-18th century.

Alloys

The main use of metallic arsenic is for alloying with copper and especially lead. Lead components in car batteries are strengthened by the presence of a few percent of arsenic. Gallium arsenide is an important semiconductor material, used in integrated circuits. Circuits made from GaAs are much faster (but also much more expensive) than those made in silicon. Unlike silicon it is direct bandgap, and so can be used in laser diodes and LEDs to directly convert electricity into light.

Military

After World War I, the United States built up a stockpile of 20000tons of lewisite ($ClCH=CHAsCl_2$), a chemical weapon that is a vesicant (blister agent) and lung irritant. The stockpile was neutralised with bleach and dumped into the Gulf of Mexico after the 1950s. During the Vietnam War the United States used Agent Blue, a mixture of sodium cacodylate and its acid form, as one of the rainbow herbicides to deprive the Vietnamese of valuable crops.

Other Uses

- Copper acetoarsenite was used as a green pigment known under many names, including 'Paris Green' and 'Emerald Green'. It caused numerous arsenic poisonings. Scheele's Green, a copper arsenate, was used in the 19th century as a colouring agent in sweets.
- Also used in bronzing and pyrotechnics.
- Up to 2% of arsenic is used in lead alloys for lead shots and bullets.
- Arsenic is added in small quantities to alpha-brass to make it dezincification resistant. This grade of brass is used to make plumbing fittings or other items that are in constant contact with water.
- Arsenic is also used for taxonomic sample preservation.
- Until recently arsenic was used in optical glass. Modern glass manufacturers, under pressure from environmentalists, have removed it, along with lead.

Biological Role

Bacteria

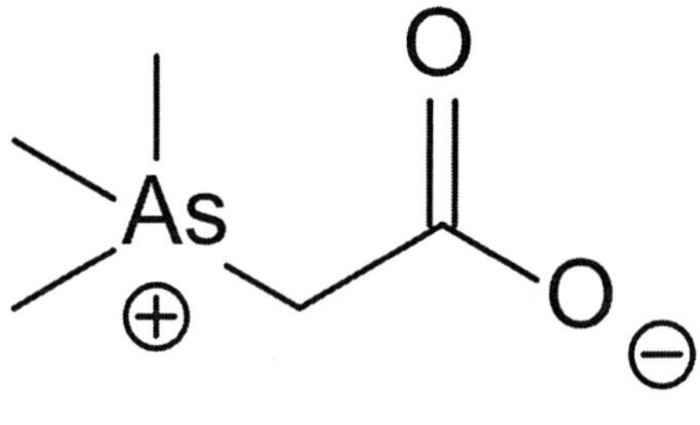

Figure 85: Arsenobetaine

Some species of bacteria obtain their energy by oxidising various fuels while reducing arsenate to arsenite. Under oxidative environmental conditions some bacteria use arsenite, which is oxidised to arsenate as fuel for their metabolism. The enzymes involved are known as arsenate reductases (Arr).

In 2008, bacteria were discovered that employ a version of photosynthesis in the absence of oxygen with arsenites as electron donors, producing arsenates (just as ordinary photosynthesis uses water as electron donor, producing molecular oxygen). Researchers conjecture that, over the course of history, these photosynthesizing organisms produced the arsenates that allowed the arsenate-reducing bacteria to thrive. One strain PHS-1 has been isolated and is related to the ã-Proteobacterium *Ectothiorhodospira shaposhnikovii*. The mechanism is unknown, but an encoded Arr enzyme may function in reverse to its known homologues.

Her.y

Arsenic has been linked to epigenetic changes that are heritable changes in gene expression that occur without changes in DNA sequence and include DNA methylation, histone modification, and RNA interference. Toxic levels of arsenic cause significant DNA hypermethylation of tumour suppressor genes p16 and p53, thus increasing risk of carcinogenesis. These epigenetic events have been observed in *in vitro* studies with human kidney cells and *in vivo* tests with rat liver cells and peripheral blood leukocytes in humans. Inductive coupled plasma mass spectrometry (ICP-MS) is used to detect precise levels of intracellular arsenic and its other bases involved in epigenetic modification of DNA. Studies investigating arsenic as an epigenetic factor will help in developing precise biomarkers of exposure and susceptibility.

The Chinese brake fern (*Pteris vittata*) hyperaccumulates arsenic present in the soil into its leaves and has a proposed use in phytoremediation.

Arsenic Reported Substituting for Phosphorus as a Building Block of Life

A NASA-funded astrobiology research team claimed on December 2, 2010 that the microbe strain GFAJ-1 of the Gammaproteobacteria (designated Halomonadaceae) group has the ability to substitute arsenic for at least part of the phosphorus in the molecules of its cells, including DNA and ATP. Bacteria from Mono Lake, a naturally arsenic-rich site in California, were cultured in an environment high in arsenic but low in phosphorus. This finding has faced strong criticism from the scientific community, many scientists have argued that there is no evidence that arsenic is actually incorporated into biomolecules. Independent confirmation of this finding has not yet been possible.

Biomethylation of Arsenic

Inorganic arsenic and its compounds, upon entering the food chain, are progressively metabolised through a process of methylation.

For example, the mold Scopulariopsis brevicaulis produce significant amounts of trimethylarsine if inorganic arsenic is present. The organic compound arsenobetaine is found in some marine foods such as fish and algae, and also in mushrooms in larger concentrations. The average person's intake is about 10–50 µg/day. Values about 1000 µg are not unusual following consumption of fish or mushrooms. But there is little danger in eating fish because this arsenic compound is nearly non-toxic.

Environmental Issues

Arsenic in Drinking Water: Widespread arsenic contamination of groundwater has led to a massive epidemic of arsenic poisoning in Bangladesh and neighbouring countries. As of this writing, 42 major incidents around the world have been reported on groundwater arsenic contamination. It is estimated that approximately 57 million people are drinking groundwater with arsenic concentrations elevated above the World Health Organisation's standard of 10 parts per billion.

However, a study of cancer rates in Taiwan suggested that significant increases in cancer mortality appear only at levels above 150 parts per billion. The arsenic in the groundwater is of natural origin, and is released from the sediment into the groundwater, owing to the anoxic conditions of the subsurface. This groundwater began to be used after local and western NGOs and the Bangladeshi government undertook a massive shallow tube well drinking-water program in the late twentieth century.

This program was designed to prevent drinking of bacteria-contaminated surface waters, but failed to test for arsenic in the groundwater. Many other countries and districts in Southeast Asia, such as Vietnam and Cambodia have geological environments conducive to generation of high-arsenic groundwaters. Arsenicosis was reported in Nakhon Si Thammarat, Thailand in 1987, and the dissolved arsenic in the Chao Phraya River is suspected of containing high levels of naturally occurring arsenic, but has not been a public health problem owing to the use of bottled water.

In the United States, arsenic is most commonly found in the ground waters of the southwest. Parts of New England, Michigan, Wisconsin, Minnesota and the Dakotas are also known to have significant concentrations of arsenic in ground water. Increased levels of skin cancer have been associated with arsenic exposure in Wisconsin, even at levels below the 10 part per billion drinking water standard.

According to a recent film funded by the US Superfund, millions of private wells have unknown arsenic levels, and in some areas of the US, over 20% of wells may contain levels that exceed established limits.

Low-level exposure to arsenic at concentrations found commonly in US drinking water compromises the initial immune response to H1N1 or swine flu infection according to NIEHS-supported scientists. The study, conducted in laboratory mice, suggests that people exposed to arsenic in their drinking water may be at increased risk for more serious illness or death in response to infection from the virus.

Epidemiological evidence from Chile shows a dose-dependent connection between chronic arsenic exposure and various forms of cancer, in particular when other risk factors, such as cigarette smoking, are present. These effects have been demonstrated to persist below 50 parts per billion.

Analysing multiple epidemiological studies on inorganic arsenic exposure suggests a small but measurable risk increase for bladder cancer at 10 parts per billion. According to Peter Ravenscroft of the Department of Geography at the University of Cambridge, roughly 80 million people worldwide consume between 10 and 50 parts per billion arsenic in their drinking water.

If they all consumed exactly 10 parts per billion arsenic in their drinking water, the previously cited multiple epidemiological study analysis would predict an additional 2,000 cases of bladder cancer alone. This represents a clear underestimate of the overall impact, since it does not include lung or skin cancer, and explicitly underestimates the exposure. Those exposed to levels of arsenic above the current WHO standard should weigh the costs and benefits of arsenic remediation.

Early (1973) evaluations of the removal of dissolved arsenic by drinking water treatment processes demonstrated that arsenic is very effectively removed by co-precipitation with either iron or aluminium oxides. The use of iron as a coagulant, in particular, was found to remove arsenic with efficiencies exceeding 90%. Several adsorptive media systems have been approved for point-of-service use in a study funded by the United States Environmental Protection Agency (U.S.EPA) and the National Science Foundation (NSF). A team of European and Indian scientists and engineers have set up six arsenic treatment plants in West Bengal based on in-situ remediation method (SAR Technology). This technology does not use any chemicals and arsenic is left as an insoluble form (+5 state) in the subterranean zone by recharging aerated water into the aquifer and thus developing an oxidation zone to support arsenic oxidising micro-organisms. This process does not produce any waste stream or sludge and is relatively cheap.

Another effective and inexpensive method to remove arsenic from contaminated well water is to sink wells 500 feet or deeper to reach purer waters. A recent 2011 study funded by the U.S. National Institute of Environmental Health Sciences' Superfund Research Program shows that deep sediments can remove arsenic and take it out of circulation. Through this process called adsorption in which arsenic sticks to the surfaces of deep sediment articles, arsenic can be naturally removed from well water.

Magnetic separations of arsenic at very low magnetic field gradients have been demonstrated in point-of-use water purification with high-surface-area and monodisperse magnetite (Fe_3O_4) nanocrystals. Using the high specific surface area of Fe_3O_4 nanocrystals the mass of waste associated with arsenic removal from water has been dramatically reduced.

Epidemiological studies have suggested a correlation between chronic consumption of drinking water contaminated with arsenic and the incidence of all leading causes of mortality. The literature provides reason to believe arsenic exposure is causative in the pathogenesis of diabetes.

Wood Preservation in the US

As of 2002, US-based industries consumed 19,600 metric tons of arsenic. Ninety percent of this was used for treatment of wood with chromated copper arsenate (CCA). In 2007, 50% of the 5,280 metric tons of consumption was still used for this purpose. In the United States, the use of arsenic in consumer products was discontinued for residential, and general consumer construction on December 31, 2003 and alternative chemicals are now used, such as Alkaline Copper Quaternary, borates, copper azole, cyproconazole, and propiconazole.

Although discontinued, this application is also one of the most concern to the general public. The vast majority of older pressure-treated wood was treated with CCA. CCA lumber is still in widespread use in many countries, and was heavily used during the latter half of the 20th century as a structural and outdoor building material. Although the use of CCA lumber was banned in many areas after studies showed that arsenic could leach out of the wood into the surrounding soil (from playground equipment, for instance), a risk is also presented by the burning of older CCA timber.

The direct or indirect ingestion of wood ash from burnt CCA lumber has caused fatalities in animals and serious poisonings in humans; the lethal human dose is approximately 20 grams of ash. Scrap CCA lumber from construction and demolition sites may be inadvertently used in commercial and domestic fires. Protocols for safe disposal of CCA lumber do not exist evenly throughout the world; there is also concern in some quarters about the widespread landfill disposal of such timber.

Mapping of Industrial Releases in the US

One tool that maps releases of arsenic to particular locations in the United States and also provides additional information about such releases is TOXMAP. TOXMAP is a Geographic Information System (GIS) from the Division of Specialised Information Services of the United States National Library of Medicine (NLM) that uses maps of the United States to help users visually explore data from the United States Environmental Protection Agency's (EPA) Toxics Release Inventory and Superfund Basic Research Programs.

TOXMAP is a resource funded by the US Federal Government. TOXMAP's chemical and environmental health information is taken from NLM's Toxicology Data Network (TOXNET) and PubMed, and from other authoritative sources.

Toxicity and Precautions

Arsenic and many of its compounds are especially potent poisons. Many water supplies close to mines are contaminated by these poisons. Arsenic disrupts ATP production through several mechanisms. At the level of the citric acid cycle, arsenic inhibits lipoic acid, which is a cofactor for pyruvate dehydrogenase; and by competing with phosphate it uncouples oxidative phosphorylation, thus inhibiting energy-linked reduction of NAD+, mitochondrial respiration and ATP synthesis.

Hydrogen peroxide production is also increased, which, it is speculated, has potential to form reactive oxygen species and oxidative stress. These metabolic interferences lead to death from multi-system organ failure, it is presumed from necrotic cell death, not apoptosis. A post mortem reveals brick-red-coloured mucosa, owing to severe haemorrhage. Although arsenic causes toxicity, it can also play a protective role.

Elemental arsenic and arsenic compounds are classified as “toxic” and “dangerous for the environment” in the European Union under directive 67/548/EEC. The International Agency for Research on Cancer (IARC) recognises arsenic and arsenic compounds as group 1 carcinogens, and the EU lists arsenic trioxide, arsenic pentoxide and arsenate salts as category 1 carcinogens.

Arsenic is known to cause arsenicosis owing to its manifestation in drinking water, “the most common species being arsenate [$HAsO_4^{2-}$; As(V)] and arsenite [H_3AsO_3 ; As(III)]”. The ability of arsenic to undergo redox conversion between As(III) and As(V) makes its availability in the environment more abundant. According to Croal, Gralnick, Malasarn and Newman, “the understanding of what stimulates As(III) oxidation and/or limits As(V) reduction is relevant for bioremediation of contaminated sites (Croal). The study of chemolithoautotrophic As(III) oxidisers and the heterotrophic As(V) reducers can help the understanding of the oxidation and/or reduction of arsenic.

Treatment of chronic arsenic poisoning is easily accomplished. British anti-lewisite (dimercaprol) is prescribed in dosages of 5 mg/kg up to 300 mg each 4 hours for the first day. Then administer the same dosage each 6 hours for the second day. Then prescribe this dosage each 8 hours for eight additional days. However the Agency for Toxic Substances and Disease Registry (ATSDR) states that the long-term effects of arsenic exposure cannot be predicted. Blood, urine, hair, and nails may be tested for arsenic; however, these tests cannot foresee possible health outcomes from the exposure. Excretion occurs in the urine and long-term exposure to arsenic has been linked to bladder and kidney cancer in addition to cancer of the liver, prostate, skin, lungs and nasal cavity.

Occupational exposure and arsenic poisoning may occur in persons working in industries involving the use of inorganic arsenic and its compounds, such as wood preservation, glass production, nonferrous metal alloys, and electronic semiconductor manufacturing. Inorganic arsenic is also found in coke oven emissions associated with the smelter industry.

Biochemical Basis of Arsenic Toxicity

The high affinity of arsenic(III) oxides for thiols is usually assigned as the cause of the high toxicity. Thiols, in the form of cysteine residues, are situated at the active sites of many important enzymes.

Arsenopyrite

Figure 86: Arsenopyrite, Panasqueira Mine, Portugal

Arsenopyrite is an iron arsenic sulfide (FeAsS). It is a hard (Mohs 5.5-6) metallic, opaque, steel grey to silver white mineral with a relatively high specific gravity of 6.1. When dissolved in nitric acid, it releases elemental sulfur. When arsenopyrite is heated, it becomes magnetic and gives off toxic fumes. With 46% arsenic content, arsenopyrite, along with orpiment, is a principal ore of arsenic. When deposits of arsenopyrite become exposed to the atmosphere, usually due to mining, the mineral will slowly oxidise, converting the arsenic into oxides that are more soluble in water, leading to acid mine drainage.

The crystal habit, hardness, density, and garlic odour when struck are diagnostic. Arsenopyrite in older literature may be referred to as *mispickel*, a name of German origin.

Arsenopyrite also can be associated with significant amounts of gold. Consequently it serves as an indicator of gold bearing reefs. Many arsenopyrite gold ores are refractory, i.e. the gold is not easily liberated from the mineral matrix.

Arsenopyrite is found in high temperature hydrothermal veins, in pegmatites, and in areas of contact metamorphism or metasomatism.

Crystallography

Figure 87: Arsenopyrite crystal from the Yaogangxian Mine, Hunan, China (size: 2.7 x 2.0 x 1.7 cm)

Arsenopyrite crystallises in the monoclinic crystal system and often shows prismatic crystal or columnar forms with striations and twinning common. Arsenopyrite may be referred to in older references as orthorhombic, but it has been shown to be monoclinic. In terms of its atomic structure, each Fe centre is linked to three As atoms and three S atoms. The material can be described as Fe^{3+} with the diatomic trianion AsS^{3-}. The connectivity of the atoms is more similar to that in marcasite than pyrite. The ion description is imperfect because the material is semiconducting and the Fe-As and Fe-S bonds are highly covalent.

Related Minerals

Various transition group metals can substitute for iron in arsenopyrite. The arsenopyrite group

Arthurite

Arthurite is a mineral with the chemical formula $CuFe_2^{3+}[(OH,O) | (AsO_4,PO_4,SO_4)]_2 \cdot 4H_2O$

In the monoclinic crystal system, it is apple-green in colour, with a Mohs hardness of 3-4, and a vitreous lustre.

Arthurite is named for the mineralogists Sir Arthur Russell and Arthur W. G. Kingsbury.

The type locality is the Hingston Down Consols mine near Gunnislake in Cornwall, in the United Kingdom, where it was discovered in 1964.

Artinite

Figure 88: Artinite sometimes forms balls of radiating, fibrous crystals. Specimen from New Idria district, California USA. Size: 9.2 x 5.2 x 1.5 cm.

Artinite is a hydrated magnesium carbonate mineral with formula: $Mg_2(CO_3)(OH)_2 \cdot 3H_2O$. It forms white silky monoclinic prismatic crystals that are often in radial arrays or encrustations. It has a Mohs hardness of 2.5 and a specific gravity of 2.

It occurs in low-temperature hydrothermal veins and in serpentinized ultramafic rocks.

It was first reported in 1902 in Lombardy, Italy. It was named for Italian mineralogist, Ettore Artini (1866-1928).

Asisite

Asisite ($Pb_7SiO_8Cl_2$) is a yellow tetragonal mineral. It is found at Kombat Mine, Kombat, Grootfontein District, Otjozondjupa Region, Namibia. It was named for a farm, Asis, which covers the mine where it was found. It was discovered in 1988.

Astrophyllite

Figure 89: Astrophyllite

Astrophyllite is a very rare, brown to golden-yellow hydrous potassium iron titanium silicate mineral. Belonging to the astrophyllite group, astrophyllite may be classed either as an inosilicate, phyllosilicate, or an intermediate between the two. It forms an isomorphous series with kupletskite, to which it is visually identical and often intimately associated. Astrophyllite is of interest primarily to scientists and collectors.

Heavy, soft and fragile, astrophyllite typically forms as *bladed*, radiating *stellate* aggregates. It is this crystal habit that gives astrophyllite its name, from the Greek words *astron* meaning "star" and *phyllon* meaning "leaf". Its great submetallic gleam and darkness contrast sharply with the light (felsic) matrix the mineral is regularly found within. Astrophyllite is usually opaque to translucent, but may be transparent in thin specimens.

As the crystals themselves possess perfect cleavage, they are typically left *in stead*, the entire aggregate often cut into slabs and polished. Owing to its limited availability and high cost, astrophyllite is seldom seen in an ornamental capacity. It is sometimes used in jewellery where it is fashioned into cabochons. Found in cavities and fissures in unusual felsic igneous rocks, astrophyllite is associated with feldspar, mica, titanite, zircon, nepheline, and aegirine. Common impurities include magnesium, aluminium, calcium, zirconium, niobium, and tantalum. It was first discovered in 1854 at its type locality; Laven Island, Norway. Interestingly, kupletskite was not known until 1956, over a hundred years later.

Astrophyllite is found in a few scarce, remote localities: Mont-Saint-Hilaire, Quebec, Canada; Pikes Peak, Colourado, USA; Narsarsuk and Kangerdluarsuk, Greenland; Brevig, Norway; and the Kola Peninsula, Russia.

Atacamite

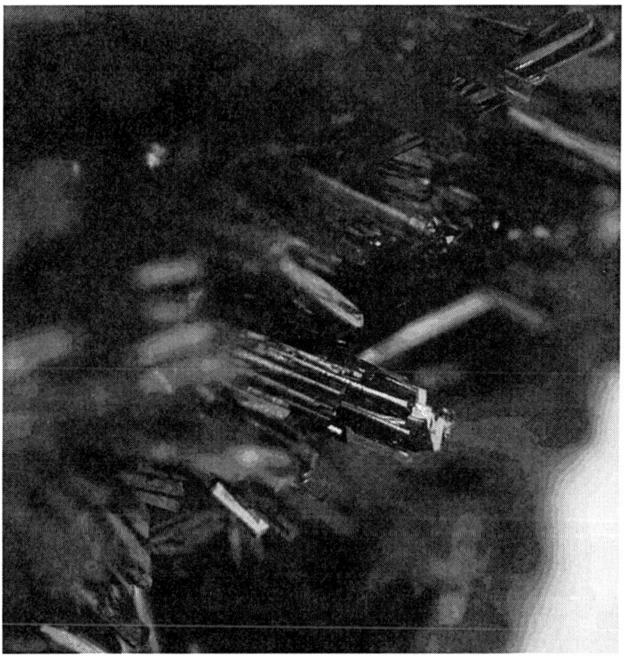

Figure 90: Atacamite prisms from Chile

Atacamite is a copper halide mineral: a copper(II) chloride hydroxide with formula $Cu_2Cl(OH)_3$.

It was first described for deposits in the Atacama Desert of Chile in 1801.

Atacamite is polymorphous with botallackite, clinoatacamite, and paratacamite. Atacamite is a comparatively rare mineral, formed from primary copper minerals in the oxidation or weathering zone of arid climates. It has also been reported from fumarole deposits, as sulfide alteration products in black smokers and as alteration of ancient bronze and copper artefacts. It occurs in association with cuprite, brochantite, linarite, caledonite, malachite, chrysocolla and its polymorphs.

Figure 91: Atacamite from Mt. Gunson, South Australia

Figure 92: Atacamite from Madagascar

Atheneite

Atheneite is a rare metallic compound commonly associated with palladium–gold deposits. It was discovered in 1974 by A.M. Clark, A.J. Criddle, and E.E. Fejer in the Minas Gerais mine in the town of Itabira, Brazil (Clark, 1974). Atheneite was also found in 1982 during a major gold rush at Serra Pelada in northern Brazil. Since then it has been found in the northern region of Para, Brazil, the northern region of Russia, and the Limpopo province of South Africa (Cabral, 2002; Trabaev, 1995).

Atheneite has a chemical structure very similar to that of graphite. Its hexagonal crystal system contributes to its metallic luster, and its colour is indicative of its chemical makeup ($(Pd,Hg)_3As$). Its name is derived from the Greek goddess Pallas Athena for its association with palladium-gold deposits. Not much is known about Atheneite because of its rarity.

Composition and Geologic Occurrence

Atheneite has the chemical formula of $(Pd,Hg)_3As$ which gives it the nomenclature of a palladium arsenide in an arsenic grouping. This mineral's composition parallels that of arsenopalladinite $Pd_8(As, Sb)_3$, isomertieite, $Pd_{11}Sb_2As_2$, and meritieite-II (Cabral, 2002). According to the American Mineralogist, the mineral was first found in concentrates from gold washings in Brazil with intergrowths of arsenopalladinite (Fleischer, 1974). Iron rich hematite was also found in intergrowths with atheneite, and is interpreted as being the product of a low temperature, hydrothermal origin.

Structure

Atheneite's space group is P6/mmm. It also has a point grouping of 6/m 2/m 2/m which concludes that its crystal system is hexagonal. Atheneite actually has a dihexagonal dipyramidal crystal form. This same complex hexagonal structure is also the structural make up of graphite. This mineral does not extinguish under crossed polars, showing bright anisotropic colours from purple-brown to dark grey (Fleischer, 1974). These birefringence colours are of the first order. The atomic structure of atheneite is very similar to that of the hexagonal structure carbons that make up graphite, however, the bonds that hold the atoms together in atheneite are much stronger than those of graphite.

Physical Properties

Atheneite, as the derivation of the name suggests, has a metallic luster like that of gold. The colour ranges from white/grey to blue/ grey tint. Its Mohs hardness scale is 4.5-5. The absolute hardness scale puts this mineral at a 48. Atheneite also has a density of 10.2 g/cm3. When placed in a reflected light, it gives off a white light with a yellowish hue and shows very weak reflectance pleochroism in oil (Fleischer, 1974).

Augelite

Figure 93: Augelite from Peru size: 3 x 3 x 2.8 cm

Augelite is an aluminium phosphate mineral with formula: $Al_2(PO_4)(OH)_3$. The shade varies from colourless to white, yellow or rose. Its crystal system is monoclinic.

It was first described by Christian Wilhelm Blomstrand for an occurrence in Västanå iron mine at Scania, Sweden in 1868.

It occurs as a product of metamorphism of phosphate bearing peraluminous sediments and in high-temperature hydrothermal ore deposits. It occurs in association with attakolite, svanbergite, lazulite, hematite, trolleite, berlinite, rutile, pyrophyllite, baryte, arsenopyrite, stannite, pyrite, andorite, cassiterite and zinkenite.

Augite

Figure 94: Augite - Muhavura volcano

Augite is a single chain inosilicate mineral, (Ca,Na)(Mg, Fe,Al)$(Si,Al)_2O_6$. The crystals are monoclinic and prismatic. Augite has two prominent cleavages, meeting at angles near 90 degrees.

Augite is a solid solution in the pyroxene group. Diopside and hedenbergite are important endmembers in augite, but augite can also contain significant aluminium, titanium, and sodium and other elements.

The calcium content of augite is limited by a miscibility gap between it and pigeonite and orthopyroxene: when occurring with either of these other pyroxenes, the calcium content of augite is a function of temperature and pressure, but mostly of temperature, and so can be useful in reconstructing temperature histories of rocks.

With declining temperature, augite may exsolve lamellae of pigeonite and/or orthopyroxene.

There is also a miscibility gap between augite and omphacite, but this gap occurs at lower temperature and is not well understood.

It is found in igneous rocks; for example, gabbro and basalt. It also occurs in relatively high-temperature metamorphic rocks such as mafic granulite.

Aurichalcite

Figure 95: Aurichalcite, 79 Mine, Banner District, Gila County, Arizona, USA

Aurichalcite is a carbonate mineral, usually found as a secondary mineral in copper and zinc deposits. Its chemical formula is $(Zn,Cu)_5(CO_3)_2(OH)_6$. The zinc to copper ratio is about 5:4.

Occurrence

Auricalcite typically occurs in the oxidised zone of copper and zinc deposits. Associated minerals include: rosasite, smithsonite, hemimorphite, hydrozincite, malachite and azurite.

It was first described in 1839 by Bottger who named the mineral for its zinc and copper content, for "mountain brass" or "mountain copper", the name of a fabulous metal. The type locality is the Loktevskoye Mine, Upper Loktevka River, Rudnyi Altai, Altaiskii Krai, Western Siberia, Russia.

Aurostibite

Aurostibite is an isometric gold antimonide mineral which is a member of the pyrite group. Aurostibite was discovered in 1952 and can be found in hydrothermal gold-quartz veins, in sulfur-deficient environments that contain other antimony minerals. The mineral can be found in Yellowknife in the Northwest Territories of Canada, and the Timiskaming District in Ontario, Canada. Antimonides are rare and are normally placed in the sulfide class by mineralogists

Autunite

Figure 96: Autunite

Autunite (hydrated calcium uranyl phosphate) with formula: $Ca(UO_2)_2(PO_4)_2 \cdot 10\text{-}12H_2O$ is a yellow - greenish fluorescent mineral with a hardness of 2 - 2½. Autunite crystallises in the tetragonal system and often occurs as tabular square crystals. Due to the moderate uranium content of 48.27% it is radioactive and also used as uranium ore. If the mineral dries out, it converts to meta-autunite-I, which can turn into meta-autunite-II after heating. These two subsequent minerals are very rare in nature. For scientific studies it is recommended to store the mineral in a sealed container to minimise the water loss. Museums are known to have covered the mineral with lacquer to avoid drying of the mineral. Autunite was discovered in 1852 near Autun, France.

Mining

Some autunite is found near the Mount Spokane, Washington. 90,000 lbs of U_3O_8 were produced from nine properties, although most of the ore came from the Daybreak Mine. The digital geologic map of Spokane county and vicinity show where the autunite is extracted. Google earth: lat: 47.885306° long:-117.175846°

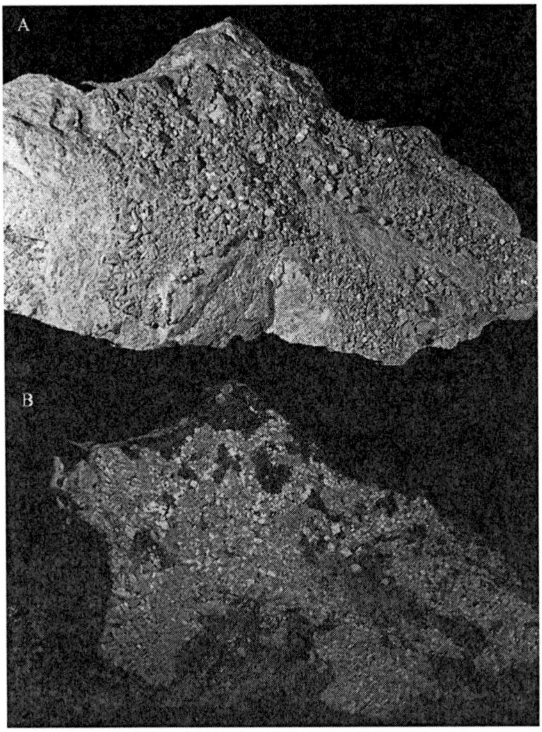

Figure 97: Autunite under UV light

Figure 98: Scaly Crystals of the secondary uranium phosphate mineral Autunite.

Azurite

Figure 99: Azurite

Azurite is a soft, deep blue copper mineral produced by weathering of copper ore deposits. It is also known as *Chessylite* after the type locality at Chessy-les-Mines near Lyon, France. The mineral has been known since ancient times, and was mentioned in Pliny the Elder's Natural History.

The blue of azurite is exceptionally deep and clear, and for that reason the mineral has tended to be associated since antiquity with the deep blue colour of low-humidity desert and winter skies. The modern English name of the mineral reflects this association, since both *azurite* and *azure* are derived via Arabic from the Persian *lazhward*, an area known for its deposits of another deep blue stone, lapis lazuli ("stone of azure").

Figure 100: Azurite, Burra Mine, South Australia

Mineralogy

Azurite crystals are monoclinic, and when large enough to be seen they appear as dark blue prismatic crystals. Azurite specimens are typically massive to nodular, and are often stalactitic in form. Specimens tend to lighten in colour over time due to weathering of the specimen surface into malachite. Azurite is soft, with a Mohs hardness of only 3.5 to 4. The specific gravity of azurite is 3.77 to 3.89. Azurite is destroyed by heat, losing carbon dioxide and water to form black, powdery copper(II) oxide. Characteristic of a carbonate, specimens effervesce upon treatment with hydrochloric acid.

Uses

Pigments:

Figure 101: Ground azurite powder for use as a pigment.

Azurite was used as a blue pigment for centuries. Depending on the degree of fineness to which it was ground, and its basic content of copper carbonate, it gave a wide range of blues. It has been known as *mountain blue* or *Armenian stone*, in addition it was formerly known as Azurro Della Magna (from Italian).

When mixed with oil it turns slightly green. When mixed with egg yolk it turns green-grey. It is also known by the names *Blue Bice* and *Blue Verditer*, though *Verditer* usually refers to a pigment made by chemical process. Older examples of azurite pigment may show a more greenish tint due to weathering into malachite. Much azurite was mislabelled *lapis lazuli*, a term applied to many blue pigments. As chemical analysis of paintings from the Middle Ages improves, azurite is being recognised as a major source of the blues used by medieval painters. True lapis lazuli was chiefly supplied from Afghanistan during the Middle Ages while azurite was a common mineral in Europe at the time. Sizable deposits were found near Lyons, France. It was mined since the 12th century in Saxony, in the silver mines located there.

Azurite was distinguished from (the much more expensive) purified natural ultramarine blue by heating (as described by Cennino D'Andrea Cennini). Ultramarine withstands heat, whereas azurite turns black (copper oxide). Gentle heating of azurite produces a deep blue pigment used in Japanese painting techniques.

Jewellery

Azurite is used occasionally as beads and as jewelry, and also as an ornamental stone. However, its softness and tendency to lose its deep blue colour as it weathers limit such uses. Heating destroys azurite easily, so all mounting of azurite specimens must be done at room temperature.

Collecting

The intense colour of azurite makes it a popular collector's stone. However, bright light, heat, and open air all tend to reduce the intensity of its colour over time. To help preserve the deep blue colour of a pristine azurite specimen, collectors should use a cool, dark, sealed storage environment similar to that of its original natural setting.

Prospecting

While not a major ore of copper itself, azurite is a good surface indicator of the presence of weathered copper sulfide ores. It is usually found in association with the chemically very similar malachite, producing a striking colour combination of deep blue and bright green that is strongly indicative of the presence of copper ores.

History

The use of azurite and malachite as copper ore indicators led indirectly to the name of the element nickel in the English language. Nickeline, a principal ore of nickel that is also known as niccolite, weathers at the surface into a green mineral (annabergite) that resembles malachite. This resemblance resulted in occasional attempts to smelt nickeline in the belief that it was copper ore, but such attempts always ended in failure due to high smelting temperatures needed to reduce nickel.

In Germany this deceptive mineral came to be known as *kupfernickel*, literally "copper demon". The Swedish alchemist Baron Axel Fredrik Cronstedt (who had been trained by Georg Brandt, the discoverer of the nickel-like metal cobalt) realised that there was probably a new metal hiding within the kupfernickel ore, and in 1751 he succeeded in smelting kupfernickel to produce a previously unknown (except in certain meteorites) silvery white, iron-like metal. Logically, Cronstedt named his new metal after the *nickel* part of *kupfernickel.* An unintended later consequence of his choice is that both Canadian and American coins worth one-twentieth of a dollar are now named after a German term for "kobolds"—that is, they are called *nickels.*

Composition

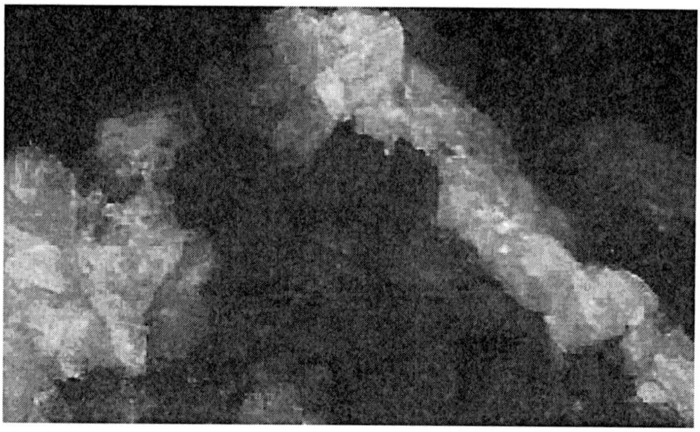

Figure 102: Fresh, unweathered stalactitic azurite crystals showing the deep blue of unaltered azurite

Azurite is one of two basic copper(II) carbonate minerals, the other being bright green malachite. Simple copper carbonate ($CuCO_3$) is not known to exist in nature. In azurite, copper(II) is linked to two different anions, carbonate and hydroxide; the compound has the formula $Cu_3(CO_3)_2(OH)_2$. The optical properties (colour, intensity) of minerals such as azurite and malachite are explained in the context of conventional electronic spectroscopy of coordination complexes. Relatively detailed descriptions are provided by ligand field theory.

Small crystals of azurite can be obtained by rapidly stirring a few drops of copper sulfate solution into a saturated solution of sodium carbonate and allowing the solution to stand overnight.

Figure 103: Azurite deposits on the interior surface of a geode

Weathering

Azurite is unstable in open air with respect to malachite, and often is pseudomorphically replaced by malachite. The weathering process involves the replacement of some the carbon dioxide (CO_2) units with water (H_2O). This changes the carbonate:hydroxide ratio of azurite from 1:1 to the 1:2 ratio of malachite:

$$2\ Cu_3(CO_3)_2(OH)_2 + H_2O \rightarrow 3\ Cu_2(CO_3)(OH)_2 + CO_2$$

From the above equation, the conversion of azurite into malachite is attributable to the low partial pressure of carbon dioxide in air. Azurite is also incompatible with aquatic media, such as saltwater aquariums.

Babingtonite

Babingtonite is a calcium iron manganese inosilicate mineral with the formula $Ca_2(Fe,Mn)FeSi_5O_{14}(OH)$. It is unusual in that iron(III) completely replaces the aluminium so typical of silicate minerals. It is a very dark green to black translucent (in thin crystals or splinters) mineral crystallising in the triclinic system with typically radial short prismatic clusters and druzy coatings.

It occurs with zeolite minerals in cavities in volcanic rocks. Babingtonite contains both iron(II) and iron(III) and shows weak magnetism. It has a Mohs hardness of 5.5 to 6 and a specific gravity of 3.3.

Figure 104: Babingtonite from Westfield, Hampden County, Massachusetts, USA

It was first described in 1824 from samples from Arendal, Aust-Agder, Norway (which is its type locality) and was named after the Irish physician and mineralogist William Babington (1757–1833).

It is the official mineral (mineral emblem) of the Commonwealth of Massachusetts.

Baddeleyite

Baddeleyite is a rare zirconium oxide mineral (ZrO_2 or zirconia), occurring in a variety of monoclinic prismatic crystal forms. It is transparent to translucent, has high indices of refraction, and ranges from colourless to yellow, green, and dark brown. Baddeleyite is a refractory mineral, with a melting point of 2700 °C. Hafnium is a substituting *impurity* and may be present in quantities ranging from 0.1 to several percents.

Figure 105: Baddeleyite from Phalaborwa, South Africa

Discovery and Occurrence

Baddeleyite was first described in 1892 from Ratnapura District, Sri Lanka. It was named after Joseph Baddeley, who described the occurrences in Sri Lanka. It has also been described from Minas Gerais and Jacupiranga, São Paulo, Brazil, and a wide variety of worldwide locations.

Baddeleyite is often found as detrital grains in gravels. Its primary occurrence is in high temperature veins and as an accessory mineral in syenite, diabase, gabbro, anorthosite, carbonatite, kimberlite and lamproite intrusions. It has also been reported from tektites, meteorites and lunar basalt. Associated minerals include ilmenite, zirkelite, apatite, magnetite, perovskite in the Jacupiranga mine, Brazil; and fluorite, nepheline, pyrochlore and allanite at Monte Somma, Italy.

Because of their refractory nature and stability under diverse conditions, baddeleyite grains, along with zircon, are used for uranium-lead radiometric age determinations.

Baotite

Baotite $Ba_4Ti_4(Ti, Nb, Fe)_4(Si_4O_{12})O_{16}Cl$ is a rare mineral recognised as having a unique four-fold silicate ring. Crystals are tetragonal, though commonly deformed to the extent of appearing monoclinic. Named for the locality of first discovery, Baotou, China, baotite has been found in hydrothermal veins and alkalic rocks in various locations around the world.

Structure

Simonov first determined the crystal structure of baotite in which (Ti, Nb, Fe)-octahedra share edges, forming chains and cross-linking at the corners forming a 4-fold screw axis in the (001); these columns are analogous to those in rutile. Four silicate tetrahedra share corners creating characteristic rings in the plane perpendicular to the c axis. The barium cation is between the rings and rutile columns, while chloride fills the void between each pair of rings. Chlorine's presence is not necessary for the structure to balance electrostatically.

Geologic Occurrence

The first sample of baotite was found with galena, pyrite, albite, aegirine and alkali amphiboles in a quartz vein cutting quartzite at the Bayan Obo Mine in China. Grains from Ravalli County, Montana were found in a thin calcite crust on masses of eschynite within a carbonatic vein. Baotite also occurs at Haast River, New Zealand as an accessory phase of carbonatite, fenite and hydrothermal fluid veins associated with a dike swarm. The dike intrusion of mantle-derived magmas can likely be attributed to the close proximity of the Alpine fault plate boundary across South Island, New Zealand.

Baryte

Figure 106: Baryte crystals on dolomite from Cumbria, England (size: 5.4×4.6 cm)

Baryte, or barite, ($BaSO_4$) is a mineral consisting of barium sulfate. The *baryte group* consists of baryte, celestine, anglesite and anhydrite. Baryte itself is generally white or colourless, and is the main source of barium. Baryte and celestine form a solid solution $(Ba,Sr)SO_4$.

Names and History

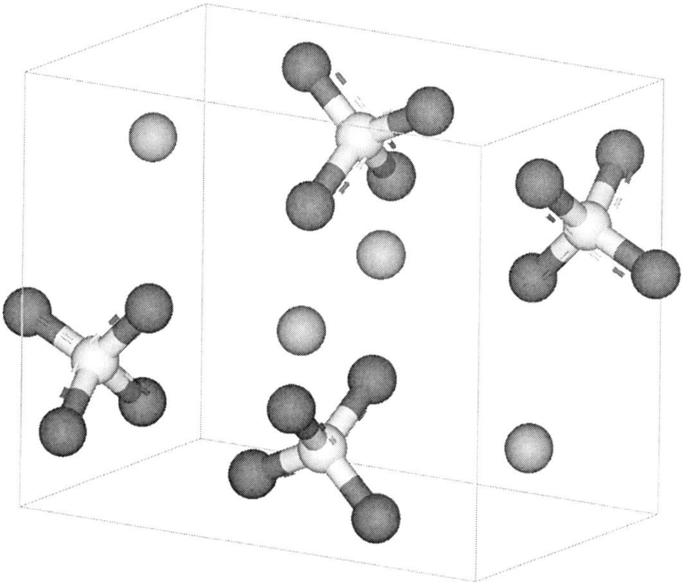

Figure 107: The unit cell of barite

The radiating form, sometimes referred to as *Bologna Stone*, attained some notoriety among alchemists for the phosphorescent specimens found in the 17th century near Bologna by Vincenzo Casciarolo.

The name baryte is derived from the Greek word heavy. The American spelling *barite* is used by the USGS and more often used in modern Scientific journals including those published by the Netherlands-based Elsevier journals. The International Mineralogical Association adopted "barite" as the official spelling when it formed in 1959, but recommended adopting the older "baryte" spelling in 1978, notably ignored by the Mineralogical Society of America. The American Petroleum Institute specification API 13/ISO 13500 which governs *baryte* for drilling purposes does not refer to any specific mineral, but rather a material that meets that specification, in practice this is usually the mineral baryte. The term "primary baryte" refers to the first marketable product, which includes crude baryte (run of mine) and the products of simple beneficiation methods, such as washing, jigging, heavy media separation, tabling, flotation. Most crude baryte requires some upgrading to minimum purity or density. Baryte that is used as an aggregate in a "heavy" cement is crushed and screened to a uniform size. Most baryte is ground to a small, uniform size before it is used as a filler or extender, an addition to industrial products, or a weighting agent in petroleum well drilling mud.

Other Names

Baryte has gone by other names such as *barytine, barytite, schwerspath, barytes, Heavy Spar,* or *tiff.*

Mineral Associations and Locations

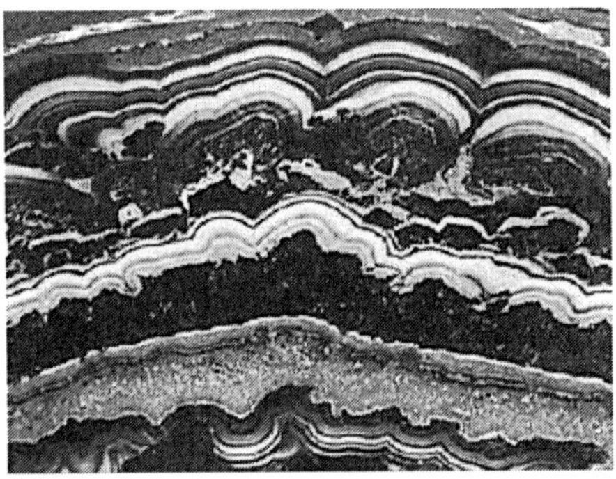

Figure 108: Baryte with Galena and Hematite from Poland

Figure 109: Large barite crystals from Nevada, USA

Figure 110: Abandoned baryte mine shaft near Aberfeldy, Perthshire, Scotland.

Baryte occurs in a large number of depositional environments, and is deposited through a large number of processes including biogenic, hydrothermal, and evaporation, among others. Baryte commonly occurs in lead-zinc veins in limestones, in hot spring deposits, and with hematite ore. It is often associated with the minerals anglesite and celestine. It has also been identified in meteorites.

In the USA, baryte has been found at locations in Cheshire, Connecticut, De Kalb, New York, Fort Wallace, New Mexico, and is quarried in Arkansas, Connecticut, Virginia, North Carolina, Georgia, Tennessee, Kentucky, Nevada, and Missouri. Localities outside the USA include Baia Sprie, Romania, Westmoreland, Cornwall, Cumberland, Derbyshire, Durham, Muirshiel (where zinc was also retracted), Perthshire, Argyllshire and Surrey in the UK, China, India, Morocco, Peru, Chile, Liberia, Turkey, Thailand, Ireland (where it was actively mined on Benbulben), Canada, Iran, Brazil, Greece, and Barberton Mountain Land, South Africa.

The major barite producers (in thousand tonnes, data for 2010) are as follows: China (3,600), India (1,000), United States (670), Morocco (460), Iran (250), Turkey (150) and Kazakhstan (100).

Uses

Some 77% worldwide is used as a weighting agent for drilling fluids in oil and gas exploration to suppress high formation pressures and prevent blowouts. As a well is drilled, the bit passes through various formations, each with different characteristics.

The deeper the hole, the more barite is needed as a percentage of the total mud mix. An additional benefit of barite is that it is non-magnetic and thus does not interfere with magnetic measurements taken in the borehole, either during logging-while-drilling or in separate drill hole logging. Barite used for drilling petroleum wells can be black, blue, brown or gray depending on the ore body.

The barite is finely ground so that at least 97% of the material, by weight, can pass through a 200-mesh (75-ìm) screen, and no more than 30%, by weight, can be less than 6 ìm diametre. The ground barite also must be dense enough so that its specific gravity is 4.2 or greater, soft enough to not damage the bearings of a tricone drill bit, chemically inert, and containing no more than 250 milligrams per kilogram of soluble alkaline salts.

Other uses are in added-value applications which include filler in paint and plastics, sound reduction in engine compartments, coat of automobile finishes for smoothness and corrosion resistance, friction products for automobiles and trucks, radiation-shielding cement, glass ceramics and medical applications (for example, a barium meal before a contrast CAT scan).

Figure 111: Baryte with Cerussite from Morocco

Baryte is supplied in a variety of forms and the price depends on the amount of processing; filler applications commanding higher prices following intense physical processing by grinding and micronising, and there are further premiums for whiteness and brightness and colour.

Historically baryte was used for the production of barium hydroxide for sugar refining, and as a white pigment for textiles, paper, and paint.

Although baryte contains a "heavy" metal (barium), it is not considered to be a toxic chemical by most governments because of its extreme insolubility.

Paleothermometry

In the deep ocean, away from continental sources of sediment, pelagic baryte crystallises out and forms a significant amount of the sediments. Since baryte has oxygen, systematics in the ä[18]O of these sediments have been used to help constrain paleotemperatures for oceanic crust. Similarly the variations in sulfur isotopes are also being exploited.

Bastnäsite

Figure 112: Bastnäsite from Burundi

The mineral bastnäsite (or bastnaesite) is one of a family of three carbonate-fluoride minerals, which includes bastnäsite-(Ce) with a formula of (Ce, La)CO_3F, bastnäsite-(La) with a formula of (La, Ce)CO_3F, and bastnäsite-(Y) with a formula of (Y, Ce)CO_3F. Most bastnäsite is bastnäsite-(Ce), and cerium is by far the most common of the rare earths in this class of minerals. Bastnäsite and the phosphate mineral monazite are the two largest sources of cerium and other rare earth elements.

Bastnäsite was first described by the Swedish chemist Wilhelm Hisinger in 1838. It is named for the Bastnäs mine near Riddarhyttan, Västermanland, Sweden. Bastnäsite also occurs as very high quality specimens at the Zagi Mountains, Pakistan. Bastnäsite occurs in alkali granite and syenite and in associated pegmatites. It also occurs in carbonatites and in associated fenites and other metasomatites.

Composition

Figure 113: Bastnäsite crystal from the Manitou District, El Paso County, Colourado, USA (size: 4.3×3.8×3.3 cm)

Bastnäsite has cerium, lanthanum and yttrium in its generalised formula but officially the mineral is divided into three minerals based on the predominant rare earth element. There is bastnäsite-(Ce) with a more accurate formula of (Ce, La)CO_3F. There is also bastnäsite-(La) with a formula of (La, Ce)CO_3F. And finally there is bastnäsite-(Y) with a formula of (Y, Ce)CO_3F. There is little difference in the three in terms of physical properties and most bastnäsite is bastnäsite-(Ce). Cerium in most natural bastnäsites usually dominates the others. Bastnäsite and the phosphate mineral monazite are the two largest sources of cerium, an important industrial metal.

Bastnäsite is closely related to the mineral series parisite. The two are both rare earth fluorocarbonates, but parisite's formula of

$Ca(Ce, La, Nd)_2(CO_3)_3F_2$ contains calcium (and a small amount of neodymium) and a different ratio of constituent ions. Parisite could be viewed as a molecule of calcite ($CaCO_3$) added to two molecules of bastnäsite. In fact, the two have been shown to alter back and forth with the addition or loss of $CaCO_3$ in natural environments.

Bastnäsite forms a series with the minerals hydroxylbastnasite-(Ce) and hydroxylbasänasite-(Nd). Th.! three are members of a substitution series that involves the possible substitution of fluorine ions for hydroxyl (OH) ion groups. Hydroxylbastnasite-(Ce) has a formula of $(Ce,La)CO_3(OH,F)$.

Name

Figure 114: Bastnäsite crystal, Zagi Mountain, Federally Administered Tribal Areas, Pakistan. Size: 1.5×1.5×0.3 cm.

Bastnäsite gets its name from its type locality, the Bastnäs Mine, Riddarhyttan, Västmanland, Sweden. Ore from the Bastnäs Mine led to the discovery of several new minerals and chemical elements by Swedish scientists such as Jöns Jakob Berzelius, Wilhelm Hisinger and Carl Gustav Mosander. Among these are the chemical elements cerium, which was described by Hisinger in 1803, and lanthanum in 1839. Hisinger, who was also the owner of the Bastnäs mine, chose to name one of the new minerals *bastnäsit* when it was first described by him in 1838.

Occurrence

Although a scarce mineral and never in great concentrations, it is widespread, and one of the more common rare-earth carbonates. Bastnäsite has been found in karst bauxite deposits in Hungary, Greece and the Balkans region.

Also found in carbonatites, a rare carbonate igneous intrusive rock, at Fen, Norway; Bayan Obo, Mongolia; Kangankunde, Malawi; Kizilcaoren, Turkey and the Mountain Pass rare earth mine in California, USA. At Mountain Pass, bastnäsite is the leading ore mineral. Some bastnäsite has been found in the unusual granites of the Langesundsfjord area, Norway; Kola Peninsula, Russia; Mont Saint-Hilaire mines, Ontario, and Thor Lake deposits, Northwest Territories, Canada. Hydrothermal sources have also been reported.

Mining History

In 1949, the huge carbonatite-hosted bastnäsite deposit was discovered at Mountain Pass, San Bernardino County, California. This discovery alerted geologists as the existence of a whole new class of rare earth deposit: the rare earth containing carbonatite.

Other examples were soon recognised, particularly in Africa and China. The exploitation of this deposit began in the mid-1960s after it had been purchased by Molycorp (Molybdenum Corporation of America). The lanthanide composition of the ore included 0.1% europium oxide, which was sorely needed by the burgeoning colour television industry, to provide the red phosphor, so as to maximise picture brightness.

The composition of the lanthanides was about 49% cerium, 33% lanthanum, 12% neodymium, and 5% praseodymium, with some samarium and gadolinium, or distinctly more lanthanum and less neodymium and heavies as compared to commercial monazite. However, the europium content was at least double that of a typical monazite.

Mountain Pass bastnäsite was the world's major source of lanthanides from the 1960s to the 1980s. Thereafter, China became increasingly important to world rare earth supply. Chinese deposits of bastnäsite include several in Sichuan Province, and the massive deposit at Bayan Obo, Inner Mongolia, which had been discovered early in the 20th century, but not exploited until much later. Bayan

Obo is currently (2008) providing the majority of the world's lanthanides. Bayan Obo bastnäsite occurs in association with monazite (plus enough magnetite to sustain one of the largest steel mills in China), and unlike carbonatite bastnäsites, is relatively closer to monazite lanthanide compositions, with the exception of its generous 0.2% content of europium.

Ore Technology

At Mountain Pass, bastnäsite ore was finely ground, and subjected to flotation to separate the bulk of the bastnäsite from the accompanying barite, calcite, and dolomite. Marketable products include each of the major intermediates of the ore dressing process: flotation concentrate, acid-washed flotation concentrate, calcined acid washed bastnäsite, and finally a cerium concentrate, which was the insoluble residue left after the calcined bastnäsite had been leached with hydrochloric acid. The lanthanides that dissolved as a result of the acid treatment were subjected to solvent extraction, to capture the europium, and purify the other individual components of the ore. A further product included a lanthanide mix, depleted of much of the cerium, and essentially all of samarium and heavier lanthanides. The calcination of bastnäsite had driven off the carbon dioxide content, leaving an oxide-fluoride, in which the cerium content had become oxidised to the less basic quadrivalent state. However, the high temperature of the calcination gave less-reactive oxide, and the use of hydrochloric acid, which can cause reduction of quadrivalent cerium, led to an incomplete separation of cerium and the trivalent lanthanides. By contrast, in China, processing of bastnäsite, after concentration, starts with heating with sulfuric acid.

Bazzite

Bazzite is a beryllium scandium cyclosilicate mineral with chemical formula: $Be_3(Sc,Fe)_2Si_6O_{18}$. It crystallises in the hexagonal crystal system typically as small blue hexagonal crystals up to 2 cm length. It has a Mohs hardness of 6.5-7 and a specific gravity of 2.77 to 2.85.

Hard to distinguish from blue beryl.

Occurs in miarolitic cavities in granite, in alpine veins and in scandium bearing granitic pegmatites.

It was first described from an occurrence in Baveno, Italy. Named after the discoverer, Alessandro E. Bazzi.

Benitoite

Figure 115: Benitoite on natrolite

Benitoite (ben-EE-toe-ite) is a rare blue barium titanium silicate mineral, found in hydrothermally altered serpentinite. Benitoite fluoresces under short wave ultraviolet light, appearing bright blue to bluish white in colour. The more rarely seen clear to white benitoite crystals fluoresce red under long-wave UV light.

It was first described in 1907 by George D. Louderback, who named it *benitoite* for its occurrence near the headwaters of the San Benito River in San Benito County, California.

Benitoite occurs in a number of sites, but gemstone quality material has only been found in California. It is California's official state gem.

Associated Minerals and Locations

Benitoite typically occurs with an unusual set of minerals, along with minerals that make up its host rock. Frequently associated minerals include: natrolite, neptunite, joaquinite, serpentine and albite.

Benitoite is a rare mineral found in very few locations including San Benito County, California, Japan and Arkansas. In the San Benito occurrence, it is found in natrolite veins within glaucophane schist within a serpentinite body. In Japan, the mineral occurs in a magnesio-riebeckite-quartz-phlogopite-albite dike cutting a serpentinite body.

In 1985 benitoite was named as the official state gem of California.

Figure 116: Blue Benitoite Crystals on white natrolite, Dallas Gem Mine, San Benito Co., California, USA

Figure 117: Benitoite crystals under UV light.

Bentorite

Bentorite is a mineral with the chemical formula $Ca_6(Cr,Al)_2(SO_4)_3(OH)_{12} \cdot 26(H_2O)$. It is coloured violet to light violet. Its crystals are hexagonal to dihexagonal dipyramidal. It is transparent and has vitreous luster. It has perfect cleavage. It is not radioactive. Bentorite is rated 2 on the Mohs Scale.

The mineral was determined by Shulamit Gross in Hatrurim Formation of Danian age along the western margin of the Dead Sea, Israel, and named for Yaakov Ben-Tor, Professor at the Hebrew

University, Jerusalem and the University of California, San Diego, California, USA, for his contributions to geology and mineralogy in Israel.

Bertrandite

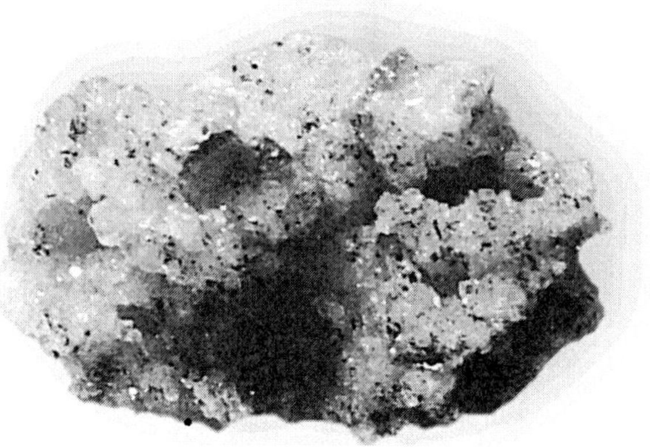

Figure 118: Bertrandite from Brazil

Bertrandite is a beryllium sorosilicate hydroxide mineral with composition: $Be_4Si_2O_7(OH)_2$. Bertrandite is a colourless to pale yellow orthorhombic mineral with a hardness of 6-7. It is commonly found in beryllium rich pegmatites and is in part an alteration of beryl. Bertrandite often occurs as a pseudomorphic replacement of beryl. It, with beryl, are ores of beryllium.

It was discovered near Nantes, France in 1883 and named after French mineralogist, Emile Bertrand (1844–1909).

Beryl

Figure 119: Three varieties of beryl: morganite, aquamarine and heliodor

The mineral beryl is a beryllium aluminium cyclosilicate with the chemical formula $Be_3Al_2(SiO_3)_6$.

The hexagonal crystals of beryl may be very small or range to several metres in size. Terminated crystals are relatively rare. Pure beryl is colourless, but it is frequently tinted by impurities; possible colours are green, blue, yellow, red, and white.

Etymology

The name beryl is derived (via Latin: *beryllus*, Old French: *beryl*, and Middle English: beril) from Greek *beryllos* which referred to a "precious blue-green colour-of-sea-water stone" and originated from Prakrit veruliya and Pali ve7uriya; ve7iru or, vi7ar, "to become pale"; ultimately from Sanskrit vaidurya-, which is of Dravidian origin, maybe from the name of Belur.

The term was later adopted for the mineral beryl more exclusively. The Late Latin word *berillus* was abbreviated as *brill-* which produced the Italian word *brillare* meaning "shine", the French word *brille* meaning "shine" and the English word *brilliance.*

Deposits

Beryl of various colours is found most commonly in granitic pegmatites, but also occurs in mica schists in the Ural Mountains, and limestone in Colombia. Beryl is often associated with tin and tungsten ore bodies.

Beryl is found in Europe in Norway, Austria, Germany, Sweden (especially morganite), Ireland and Russia, as well as Brazil, Colombia, Madagascar, Mozambique, South Africa, the United States, and Zambia. U.S. beryl locations are in California, Colourado, Connecticut, Idaho, Maine, New Hampshire, North Carolina, South Dakota and Utah.

New England's pegmatites have produced some of the largest beryls found, including one massive crystal from the Bumpus Quarry in Albany, Maine with dimensions 5.5 m by 1.2 m (18 ft by 4 ft) with a mass of around 18 metric tons; it is New Hampshire's state mineral.

As of 1999, the largest known crystal of any mineral in the world is a crystal of beryl from Madagascar, 18 metres long and 3.5 metres in diametre.

Varieties

Aquamarine and Maxixe

Figure 120: Aquamarine

Aquamarine (from Latin: *aqua marina*, "water of the sea") is a blue or turquoise variety of beryl. It occurs at most localities which yield ordinary beryl, some of the finest coming from Russia. The gem-gravel placer deposits of Sri Lanka contain aquamarine. Clear yellow beryl, such as that occurring in Brazil, is sometimes called aquamarine chrysolite.

When corundum presents the bluish tint of typical aquamarine, it is often termed Oriental aquamarine. The deep blue version of aquamarine is called *maxixe*. Its colour fades to white when exposed to sunlight or is subjected to heat treatment, though the colour returns with irradiation.

The pale blue colour of aquamarine is attributed to Fe^{2+}. The Fe^{3+} ions produce golden-yellow colour, and when both Fe^{2+} and Fe^{3+} are present, the colour is a darker blue as in maxixe. Decolouration of maxixe by light or heat thus may be due to the charge transfer Fe^{3+} and Fe^{2+}. Dark-blue maxixe colour can be produced in green, pink or yellow beryl by irradiating it with high-energy particles (gamma rays, neutrons or even X-rays).

In the United States, aquamarines can be found at the summit of Mt. Antero in the Sawatch Range in central Colourado. In Wyoming, aquamarine has been discovered in the Big Horn Mountains, near Powder River Pass. In Brazil, there are mines in the states of Minas Gerais, Espírito Santo, and Bahia, and minorly in Rio Grande do

Norte. The Mines of Colombia, Zambia, Madagascar, Malawi, Tanzania and Kenya also produce aquamarine.

The largest aquamarine of gemstone quality ever mined was found in Marambaia, Minas Gerais, Brazil, in 1910. It weighed over 110 kg, and its dimensions were 48.5 cm (19 in) long and 42 cm (17 in) in diametre.

Emerald

Figure 121: Faceted emerald gemstones

Figure 122: Rough emerald on matrix

Emerald refers to green beryl, coloured by trace amounts of chromium and sometimes vanadium.

The word "emerald" comes (via Middle English: Emeraude, imported from Old French: Ésmeraude and Medieval Latin: Esmaraldus) from Latin *smaragdus* from Greek *smaragdos* – green gem. Most emeralds are highly included, so their brittleness (resistance to breakage) is classified as generally poor.

Emeralds in antiquity were mined by the Egyptians and in Austria, as well as Swat in northern Pakistan. A rare type of emerald known as a trapiche emerald is occasionally found in the mines of Colombia. A trapiche emerald exhibits a "star" pattern; it has raylike spokes of dark carbon impurities that give the emerald a six-pointed radial pattern. It is named for the *trapiche*, a grinding wheel used to process sugarcane in the region.

Colombian emeralds are generally the most prized due to their transparency and fire. Some of the most rare emeralds come from three main emerald mining areas in Colombia: Muzo, Coscuez, and Chivor.

Fine emeralds are also found in other countries, such as Zambia, Brazil, Zimbabwe, Madagascar, Pakistan, India, Afghanistan and Russia. In the US, emeralds can be found in Hiddenite, North Carolina. In 1998, emeralds were discovered in the Yukon.

Emerald is a rare and valuable gemstone and, as such, it has provided the incentive for developing synthetic emeralds. Both hydrothermal and *flux-growth* synthetics have been produced. The first commercially successful emerald synthesis process was that of Carroll Chatham. The other large producer of flux emeralds was Pierre Gilson Sr., which has been on the market since 1964. Gilson's emeralds are usually grown on natural colourless beryl seeds which become coated on both sides. Growth occurs at the rate of 1 mm per month, a typical seven-month growth run producing emerald crystals of 7 mm of thickness. The green colour of emeralds is attributed to presence of Cr^{3+} ions.

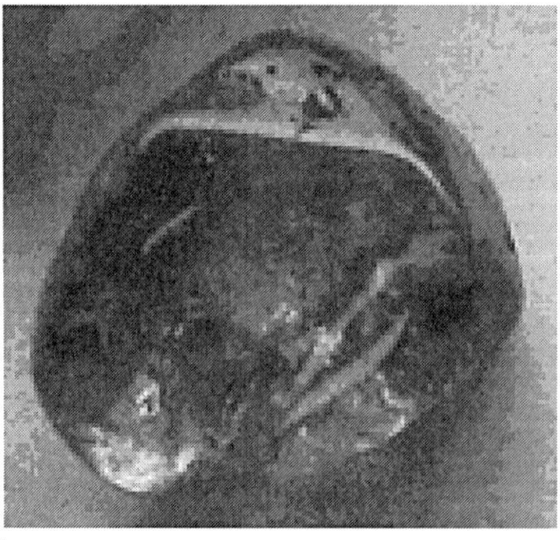

Figure 123: Heliodor

Golden Beryl and Heliodor

Figure 124: Golden beryl

Golden beryl can range in colours from pale yellow to a brilliant gold. Unlike emerald, golden beryl has very few flaws. The term "golden beryl" is sometimes synonymous with *heliodor* but golden beryl refers to pure yellow or golden yellow shades, while heliodor refers to the greenish-yellow shades. The golden yellow colour is attributed to Fe^{3+} ions. Both golden beryl and heliodor are used as gems. Probably the largest cut golden beryl is the flawless 2054 carat stone on display in the Hall of Gems, Washington, D.C.

Goshenite

Colourless beryl is called *goshenite*. The name originates from Goshen, Massachusetts where it was originally discovered. Since all these colour varieties are caused by impurities and pure beryl is colourless, it might be tempting to assume that goshenite is the purest variety of beryl. However, there are several elements that can act as inhibitors to colour in beryl and so this assumption may not always be true. The name goshenite has been said to be on its way to extinction and yet it is still commonly used in the gemstone markets. Goshenite is found to some extent in almost all beryl localities. In the past, goshenite was used for manufacturing eyeglasses and lenses owing to its transparency. Nowadays, it is most commonly used for gemstone purposes and also considered as a source of beryllium.

Figure 125: Goshenite

The gem value of goshenite is relatively low. However, goshenite can be coloured yellow, green, pink, blue and in intermediate colours by irradiating it with high-energy particles. The resulting colour depends on the content of Ca, Sc, Ti, V, Fe, and Co impurities.

Morganite

Figure 126: Morganite

Morganite, also known as "pink beryl", "rose beryl", "pink emerald", and "cesian (or *caesian*) beryl", is a rare light pink to rose-coloured gem-quality variety of beryl. Orange/yellow varieties of morganite can also be found, and colour banding is common. It can be routinely heat treated to remove patches of yellow and is occasionally treated by irradiation to improve its colour. The pink colour of morganite is attributed to Mn^{2+} ions.

Pink beryl of fine colour and good sizes was first discovered on an island on the coast of Madagascar in 1910. It was also known, with other gemstone minerals, such as tourmaline and kunzite, at Pala, California. In December 1910, the New York Academy of Sciences named the pink variety of beryl "morganite" after financier J. P. Morgan.

On October 7, 1989, one of the largest gem morganite specimens ever uncovered, eventually called "The Rose of Maine," was found at the Bennett Quarry in Buckfield, Maine, USA. The crystal, originally somewhat orange in hue, was 23 cm (9 in) long and about 30 cm (12 in) across, and weighed (along with its matrix) just over 50 lbs (23 kg).

Figure 127: Red beryl

Red Beryl

Red beryl (also known as "red emerald" or "scarlet emerald") is a red variety of beryl. It was first described in 1904 for an occurrence, its type locality, at Maynard's Claim (Pismire Knolls), Thomas Range, Juab County, Utah. The old synonym "bixbite" is deprecated from the CIBJO, because of the risk of confusion with the mineral bixbyite (also named after the mineralogist Maynard Bixby). The dark red colour is attributed to Mn^{3+} ions.

Red beryl is very rare and has only been reported from a handful of locations including: Wah Wah Mountains, Beaver County, Utah; Paramount Canyon, Sierra County, New Mexico; Round Mountain, Sierra County, New Mexico; and Juab County, Utah.

The greatest concentration of gem-grade red beryl comes from the Violet Claim in the Wah Wah Mountains of mid-western Utah, discovered in 1958 by Lamar Hodges, of Fillmore, Utah, while he was prospecting for uranium.

Prices for top quality natural red beryl can be as high as $10,000 per carat for faceted stones. Red beryl has been known to be confused with pezzottaite, also known as raspberry beryl or “raspberyl”, a gemstone that has been found in Madagascar and now Afghanistan – although cut gems of the two varieties can be distinguished from their difference in refractive index.

While gem beryls are ordinarily found in pegmatites and certain metamorphic rocks, red beryl occurs in topaz-bearing rhyolites. It formed by crystallising under low pressure and high temperature from a pneumatolitic phase along fractures or within near-surface miarolitic cavities of the rhyolite. Associated minerals include bixbyite, quartz, orthoclase, topaz, spessartine, pseudobrookite and hematite.

Beryllonite

Figure 128: Beryllonite (whitish) with tourmaline variety indicolite (blue) from Paprok, Nuristan Province, Afghanistan. Specimen size 4 cm

Beryllonite is a rare sodium beryllium phosphate mineral with formula $NaBePO_4$. The tabular to prismatic monoclinic crystals vary from colourless to white or pale yellowish, and are transparent with a vitreous lustre. Twinning is common and occurs in several forms. It exhibits perfect cleavage in one direction. The hardness is 5.5 to 6 and the specific gravity is 2.8. Refractive indices are $n\alpha = 1.552$, $n\beta = 1.558$ and nã = 1.561. A few crystals have been cut and faceted, but, as the refractive index is no higher than that of quartz, they do not make very brilliant gemstones.

It occurs as a secondary beryllium mineral in granitic and alkalic pegmatites. It was first described from complex crystals and as broken fragments in the disintegrated material of a granitic vein at Stoneham, Oxford County, Maine where it is associated with feldspar, smoky quartz, beryl and columbite. It was discovered by James Dwight Dana in 1888, and named beryllonite for its beryllium content.

Biotite

Figure 129: Thin tabular Biotite aggregate (Image width: 2.5 mm)

Biotite is a common phyllosilicate mineral within the mica group, with the approximate chemical formula $K(Mg,Fe)_3AlSi_3O_{10}(F,OH)_2$. More generally, it refers to the dark mica series, primarily a solid-solution series between the iron-endmember annite, and the magnesium-endmember phlogopite; more aluminous endmembers include siderophyllite. Biotite was named by J.F.L. Hausmann in 1847 in honour of the French physicist Jean-Baptiste Biot, who, in 1816, researched the optical properties of mica, discovering many unique properties.

Biotite is a sheet silicate. Iron, magnesium, aluminium, silicon, oxygen, and hydrogen form sheets that are weakly bound together by potassium ions. It is sometimes called "iron mica" because it is more iron-rich than phlogopite. It is also sometimes called "black mica" as opposed to "white mica" (muscovite) – both form in some rocks, in some instances side-by-side.

Properties

Like other mica minerals, biotite has a highly perfect basal cleavage, and consists of flexible sheets, or lamellae, which easily flake off. It has a monoclinic crystal system, with tabular to prismatic crystals with an obvious pinacoid termination.

It has four prism faces and two pinacoid faces to form a pseudohexagonal crystal. Although not easily seen because of the cleavage and sheets, fracture is uneven. It appears greenish to brown or black, and even yellow when weathered. It can be transparent to opaque, has a vitreous to pearly luster, and a grey-white streak. When biotite is found in large chunks, they are called "books" because it resembles a book with pages of many sheets.

Under cross polarised light biotite can generally be identified by the gnarled bird's eye extinction.

Occurrence

Biotite is found in a wide variety of igneous and metamorphic rocks. For instance, biotite occurs in the lava of Mount Vesuvius and in the Monzoni intrusive complex of the western Dolomites. It is an essential phenocryst in some varieties of lamprophyre. Biotite is occasionally found in large cleavable crystals, especially in pegmatite veins, as in New England, Virginia and North Carolina. Other notable occurrences include Bancroft and Sudbury, Ontario. It is an essential constituent of many metamorphic schists, and it forms in suitable compositions over a wide range of pressure and temperature.

The largest documented single crystals of biotite were approximately 7 m^2 (75 sq ft) sheets found in Iveland, Norway.

Uses

Figure 130: Biotite: Topotype deposit

Biotite is used extensively to constrain ages of rocks, by either potassium-argon dating or argon-argon dating. Because argon escapes readily from the biotite crystal structure at high temperatures, these methods may provide only minimum ages for many rocks. Biotite is also useful in assessing temperature histories of metamorphic rocks, because the partitioning of iron and magnesium between biotite and garnet is sensitive to temperature.

Birnessite

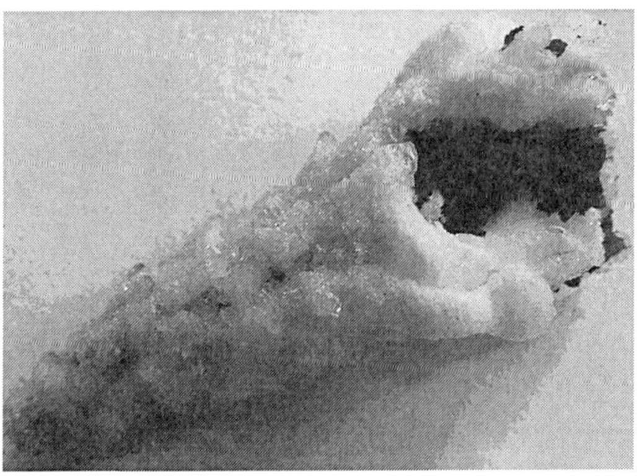

Figure 131: Birnessite

Birnessite $(Na_{0.3}Ca_{0.1}K_{0.1})(Mn^{4+},Mn^{3+})_2O_4 \cdot 1.5\ H_2O$ is an oxide mineral of manganese along with calcium, potassium and sodium. It has a dark brown to black colour with a submetallic luster. It is also very soft, with a Mohs hardness of 1.5. Birnessite is formed by precipitation in lakes, oceans and groundwater and is a major component of desert varnish and deep sea manganese nodules.

History

It was first described in 1956 and named for an occurrence in Birness, Aberdeenshire, Scotland. Birnessite is found as an oxidation product of several other minerals, including rhodonite, rhodochrosite, and as a weathering product of franklinite-wilminite ore. It has also been found as a coating along joint planes and fractures within a trachyte sill. However, it has been most commonly seen as a constituent of oceanic nodules of manganese.

A recent study found that the mineral is able to break down prions via oxidation. How well this process works outside the laboratory is unclear.

Geologic Occurrence

Levinson noted the presence of Birnessite in one or more mines from the region around Zacatecas, Mexico, while other notations have been made in Canada, at Cummington, Massachusetts, from the aforementioned nodules in both the Atlantic and Pacific Oceans, from the Tachaki Mine in Japan, the Treburland Mine in Cornwall, England, and from a bog in Norway.

Composition and Structure

Birnessite is a phyllomanganate, which is a type of ferroalloy that contains a high proportion of manganese. While natural forms generally contain foreign ions (i.e. Na, Ca, K) they are considered non-essential and synthetic forms of the mineral can be produced without them. However, most of the synthetic versions of the mineral undergo significant water loss at temperatures below 100°C. Its structure is thought to be similar to that of chalcophanite, and has been modelled as such by Burns. The structure itself consists of sheets of water molecules found between sheets of edge-sharing molecules of MnO_6 octahedra, and repeated on an average of every 7.2 Å, doing so along the c-axis. Of the six octahedral sites in the MnO_6 octahedral layer, one is left unoccupied; Mn^{2+} and Mn^{3+} lie above each vacant slot on the octahedral. These Mn ions are low-valence, and associate with O, in both the octahedral and in the water sheets.

Gypsum

Figure 132: Fibrous Gypsum Selenite showing its translucent property

Gypsum is a very soft sulfate mineral composed of calcium sulfate dihydrate, with the chemical formula $CaSO_4 \cdot 2H_2O$. It is found in alabaster, a decorative stone used in Ancient Egypt. It is the second softest mineral on the Mohs Hardness Scale. It forms as an evaporite mineral and as a hydration product of anhydrite.

Etymology and History

Gypsum was known in Old English as *sp¿rstân, spear stone,* referring to its crystalline projections. (Thus, the word spar in mineralogy is by way of comparison to gypsum, referring to any non-ore mineral or crystal that forms in spearlike projections.) In the early nineteenth century it was regarded as an almost miraculous fertilizer. American farmers were so anxious to acquire it that a lively smuggling trade with Nova Scotia evolved, resulting in the so-called "Plaster War" of 1812.

Physical Properties

Gypsum is moderately water-soluble (~2.0–2.5 g/L at 25 °C) and, in contrast to most other salts, it exhibits a retrograde solubility, becoming less soluble at higher temperatures. When the crystal lattice is heated it loses liquid water molecules to evaporation and thus gains solidity. As for anhydrite, its solubility in saline solutions and in brines is also strongly dependent on NaCl concentration.

Gypsum crystals are found to contain anion water and hydrogen bonding.

Crystal Varieties

Gypsum occurs in nature as flattened and often twinned crystals and transparent cleavable masses called selenite. Selenite contains no significant selenium; rather both substances were named from the Ancient Greek word for the Moon.

Selenite may also occur in a silky, fibrous form, in which case it is commonly called *satin spar*. Finally it may also be granular or quite compact.

In hand-sized samples, it can be anywhere from transparent to opaque. A very fine-grained white or lightly tinted variety of gypsum is called alabaster, and is prized for ornamental work of various sorts. In arid areas, gypsum can occur in a flower-like form typically opaque with embedded sand grains called desert rose. Gypsum forms some of the largest crystals found in nature, up to 11 metres long, in the form of selenite.

Occurrence

Figure 133: Veins of gypsum in the silts/marls of the Tea Green and Grey Marls, Blue Anchor, Somerset, UK

Gypsum is a common mineral, with thick and extensive evaporite beds in association with sedimentary rocks. Deposits are known to occur in strata from as far back as the Archaean eon. Gypsum is deposited from lake and sea water, as well as in hot springs, from volcanic vapours, and sulfate solutions in veins.

Hydrothermal anhydrite in veins is commonly hydrated to gypsum by groundwater in near surface exposures. It is often associated with the minerals halite and sulfur.

Figure 134: Veins of gypsum in the Chugwater Group, Wyoming

Because gypsum dissolves over time in water, gypsum is rarely found in the form of sand. However, the unique conditions of the White Sands National Monument in the US state of New Mexico have created a 710 km^2 (270 sq mi) expanse of white gypsum sand, enough to supply the construction industry with drywall for 1,000 years.

Commercial exploitation of the area, strongly opposed by area residents, was permanently prevented in 1933 when president Herbert Hoover declared the gypsum dunes a protected national monument.

Gypsum is also formed as a by-product of sulfide oxidation, amongst others by pyrite oxidation, when the sulfuric acid generated reacts with calcium carbonate.

Its presence indicates oxidising conditions. Under reducing conditions, the sulfates it contains can be reduced back to sulfide by sulfate reducing bacteria.

Orbital pictures from the Mars Reconnaissance Orbiter (MRO) indicate the existence of gypsum dunes in the northern polar region of Mars.

Mining

Figure 135: Golden Gypsum crystals from Winnipeg.

Figure 136: Gypsum sand from White Sands National Monument, New Mexico.

Commercial quantities of gypsum are found in the cities of Araripina and Grajaú, Brazil, Pakistan, Jamaica, Iran (world's second largest producer), Thailand, Spain (the main producer in Europe), Germany, Italy, England, Ireland, in British Columbia, Manitoba, Ontario, Nova Scotia and Newfoundland in Canada, and in New York, Michigan, Indiana, Texas (in the Palo Duro Canyon), Iowa, Kansas,

Oklahoma, Arizona, New Mexico, Colourado, Utah, Arkansas and Nevada in the United States. There is also a large open pit quarry located at Plaster City, California, in Imperial County, and in East Kutai, Kalimantan. Several small mines also exist in places like Kalannie in Western Australia where gypsum is sold to private buyers for changing the pH levels of soil for agricultural purposes.

Crystals of gypsum up to 11 metres (36 ft) long have been found in the caves of the Naica Mine of Chihuahua, Mexico. The crystals thrived in the cave's extremely rare and stable natural environment. Temperatures stayed at 58 °C (136 °F), and the cave was filled with mineral-rich water that drove the crystals' growth. The largest of those crystals weighs 55 short tons (50,000 kg) and is around 500,000 years old.

Synthesis

Synthetic gypsum is recovered via flue gas desulfurization at some coal-fired electric power plants. It can be used interchangeably with natural gypsum in some applications.

Gypsum also precipitates onto brackish water membranes, a phenomenon known as mineral salt scaling, such as during brackish water desalination of water with high concentrations of calcium and sulfate. Scaling decreases membrane life and productivity. This is one of the main obstacles in brackish water membrane desalination processes, such as reverse osmosis or nanofiltration. Other forms of scaling such as calcite scaling, depending on the water source, can also be important considerations in distillation as well as in heat exchangers where either the salt solubility or salt concentration can change rapidly.

Uses of Gypsum

Gypsum is used in a wide variety of applications:

- Gypsum board primarily used as a finish for walls and ceilings; known in construction as drywall.
- Plaster ingredient.
- Fertilizer and soil conditioner. In the late 18th and early 19th century, Nova Scotia gypsum, often referred to as plaster, was a highly sought fertilizer for wheat fields in the United States. It is also used in ameliorating sodic soils.

- A binder in fast-dry tennis court clay.
- Plaster of Paris (surgical splints; casting moulds; modelling).
- A wood substitute in the ancient world; for example, when wood became scarce due to deforestation on Bronze Age Crete, gypsum was employed in building construction at locations where wood was previously used.
- A tofu (soy bean curd) coagulant, making it ultimately a major source of dietary calcium, especially in Asian cultures which traditionally use few dairy products.
- Adding hardness to water used for homebrewing.
- A component of Portland cement used to prevent flash setting of concrete.
- Soil/water potential monitoring (soil moisture tension).
- A common ingredient in making mead.
- In the medieval period it was mixed, by scribes and illuminators, with lead carbonate (powdered white lead) to make gesso which was applied to illuminated letters and gilded with gold in illuminated manuscripts.
- In foot creams, shampoos and many other hair products.
- A medicinal agent in traditional Chinese medicine called *Shi Gao*.
- Impression plasters in dentistry
- Unusual gypsum specimens from around the world

Figure 137: Green gypsum crystals from Pernatty Lagoon, Mt Gunson, South Australia. Its green colour is due to presence of copper ions

Figure 138: Desert rose, 47 cm long

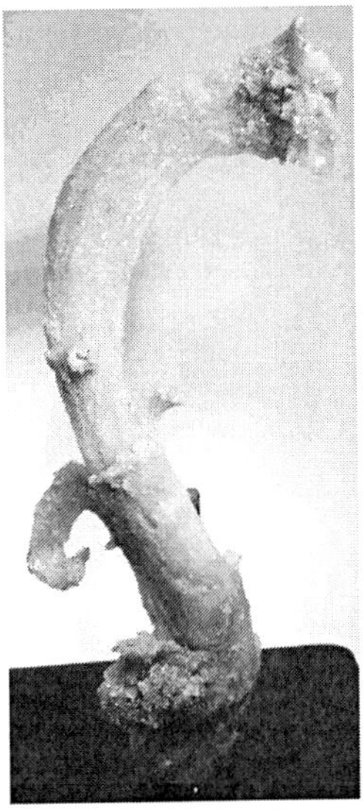

Figure 139: Classic "ramshorn" gypsum from Santa Eulalia, Chihuahua, Mexico. Size 7.5×4.3×3.8 cm

Figure 140: Selenite from the Houston Museum of Natural Science

Figure 141: Gypsum with inclusions of green atacamite from Peru

Figure 142: Gypsum with crystalline native copper inside

Figure 143: Bright, cherry-red gypsum crystals 2.5 cm in height coloured by rich inclusions of the rare mineral botryogen

Figure 144: Fiery orange gypsum crystals

Figure 145: Unusual selenite gypsum from the Red River, Winnipeg, Manitoba, Canada.

Kaolinite

Kaolinite is a clay mineral, part of the group of industrial minerals, with the chemical composition $Al_2Si_2O_5(OH)_4$. It is a layered silicate mineral, with one tetrahedral sheet linked through oxygen atoms to one octahedral sheet of alumina octahedra. Rocks that are rich in kaolinite are known as kaolin or china clay.

The name is derived from *Kao-ling* (Chinese: Øšu–/Øš°]; pinyin: *Gaoling*), a town near Jingdezhen, Jiangxi province, China. The name entered English in 1727 from the French version of the word: "kaolin", following Francois Xavier d'Entrecolles's reports from Jingdezhen.

Kaolinite has a low shrink-swell capacity and a low cation exchange capacity (1-15 meq/100g). It is a soft, earthy, usually white mineral (dioctahedral phyllosilicate clay), produced by the chemical weathering of aluminium silicate minerals like feldspar. In many parts of the world, it is coloured pink-orange-red by iron oxide, giving it a distinct rust hue. Lighter concentrations yield white, yellow or light orange colours. Alternating layers are sometimes found, as at Providence Canyon State Park in Georgia, USA.

Figure 146: Kaolin

Structural Transformations

Kaolin-type clays undergo a series of phase transformations upon thermal treatment in air at atmospheric pressure. Endothermic dehydroxylation (or alternatively, dehydration) begins at 550–600 °C to produce disordered metakaolin, $Al_2Si_2O_7$, but continuous hydroxyl loss (-OH) is observed up to 900 °C and has been attributed to gradual oxolation of the metakaolin. Because of historic disagreement concerning the nature of the metakaolin phase, extensive research has led to general consensus that metakaolin is not a simple mixture of amorphous silica (SiO_2) and alumina (Al_2O_3), but rather a complex amorphous structure that retains some longer-range order (but not strictly crystalline) due to stacking of its hexagonal layers.

$$2\ Al_2Si_2O_5(OH)_4 \rightarrow 2\ Al_2Si_2O_7 + 4\ H_2O.$$

Further heating to 925–950 °C converts metakaolin to a defect aluminium-silicon spinel, $Si_3Al_4O_{12}$, which is sometimes also referred to as a gamma-alumina type structure:

$$2\ Al_2Si_2O_7 \rightarrow Si_3Al_4O_{12} + SiO_2.$$

Upon calcination to ~1050 °C, the spinel phase ($Si_3Al_4O_{12}$) nucleates and transforms to mullite, 3 $Al_2O_3 \cdot 2 SiO_2$, and highly crystalline cristobalite, SiO_2:

$$3\ Si_3Al_4O_{12} \rightarrow 2\ Si_2Al_6O_{13} + 5\ SiO_2.$$

Occurrence

Figure 147: A kaolin mine in Ruse Province, Bulgaria

Kaolinite is one of the most common minerals; it is mined, as kaolin, in Brazil, Bulgaria, France, United Kingdom, Iran, Germany, India, Australia, Korea, the People's Republic of China, the Czech Republic and the United States.

[:] Predominance in tropical soils

Kaolinite clay occurs in abundance in soils that have formed from the chemical weathering of rocks in hot, moist climates—for example in tropical rainforest areas. Comparing soils along a gradient towards progressively cooler or drier climates, the proportion of kaolinite decreases, while the proportion of other clay minerals such as illite (in cooler climates) or smectite (in drier climates) increases. Such climatically-related differences in clay mineral content are often used to infer changes in climates in the geological past, where ancient soils have been buried and preserved.

In thc *Institut National pour l'Etude Agronomique au Congo Belge* (INEAC) classification system, soils in which the clay fraction is predominantly kaolinite are called *kaolisol* (from kaolin and soil).

Uses

Kaolin is used in ceramics, medicine, coated paper, as a food additive in toothpaste, as a light diffusing material in white incandescent light bulbs, and in cosmetics. It is generally the main component in porcelain.

It is also used in paint to extend titanium dioxide (TiO_2) and modify gloss levels; in rubber for semi-reinforcing properties; and in adhesives to modify rheology. Kaolin was long used in the production of common smoking pipes in Europe and Asia.

The largest use is in the production of paper, including ensuring the gloss on some grades of paper. Commercial grades of kaolin are supplied and transported as dry powder, semi-dry noodle or as liquid slurry.

Kaolinite can contain very small traces of uranium and thorium, and is therefore useful in radiological dating. While a single magazine made using kaolin does not contain enough radioactive material to be detected by a security-oriented monitor, this could result in the possibility of a truckload of glossy paper occasionally tripping an overly-sensitive radiation monitor.

Kaolinite has also seen some use in organic farming, as a spray applied to crops to deter insect damage, and in the case of apples, to prevent sun scald.

In April 2008, the US Naval Medical Research Institute announced the successful use of a kaolinite-derived aluminosilicate nanoparticle infusion in traditional gauze, known commercially as QuikClot Combat Gauze.

When heated to between 650 and 900 °C kaolinite dehydroxylates to form metakaolin. According to the American National Precast Concrete Association this is a supplementary cementitious material (SCM). When added to a concrete mix, metakaolin affects the acceleration of Portland cement hydration when replacing Portland cement by 20 percent by weight.

In ceramics applications, the formula is typically written in terms of oxides, thus the formula for kaolinite is $Al_2O_3 \cdot 2SiO_2 \cdot 2H_2O$

Using the cement chemist notation this can be even written as AS_2H_2, with the oxides represented as $A = Al_2O_3$, $S = SiO_2$, $H = H_2O$.

Kaolinite was also used as filler in Edison Diamond Discs.

Medicinal and Culinary Uses

A folk medicine use is to soothe an upset stomach, similar to the way parrots (and later, humans) in South America originally used it. More recently, industrially-produced kaolinite preparations were formerly common for treatment of diarrhea; the most common of these was Kaopectate, which abandoned the use of kaolin in favour of attapulgite and then (in the United States) bismuth subsalicylate.

Figure 148: Kaolin

In Africa, kaolin is sometimes known as *kalaba* (in Gabon and Cameroon, *calaba*, and *calabachop* (in Equatorial Guinea).

It is used for facial masks or soap and is eaten for pleasure or to suppress hunger, a practice known as geophagy. Consumption is greater among women, especially during pregnancy.

This practice is also seen among black women in the Southern United States, especially Georgia. There, the kaolin is called *white dirt*, *chalk* or *white clay*.

Kyanite

Figure 149: Kyanite

Kyanite, whose name derives from the Greek word *kuanos* sometimes referred to as "kyanos", meaning deep blue, is a typically blue silicate mineral, commonly found in aluminium-rich metamorphic pegmatites and/or sedimentary rock. Kyanite in metamorphic rocks generally indicates pressures higher than 4 kilobars. Although potentially stable at lower pressure and low temperature, the activity of water is usually high enough under such conditions that it is replaced by hydrous aluminosilicates such as muscovite, pyrophyllite, or kaolinite. Kyanite is also known as disthene, rhaeticite and cyanite.

Kyanite is a member of the aluminosilicate series, which also includes the polymorph andalusite and the polymorph sillimanite. Kyanite is strongly anisotropic, in that its hardness varies depending on its crystallographic direction. In Kyanite, this anisotropism can be considered an identifying characteristic.

At temperatures above 1100 °C kyanite decomposes into mullite and vitreous silica via the following reaction: $3(Al_2O_3 \cdot SiO_2) \rightarrow 3Al_2O_3 \cdot 2SiO_2 + SiO_2$. This transformation results in an expansion.

Uses of Kyanite

Kyanite is used primarily in refractory and ceramic products, including porcelain plumbing fixtures and dishware. It is also used in electronics, electrical insulators and abrasives.

Kyanite has been used as a semiprecious gemstone, which may display cat's eye chatoyancy, though this use is limited by its anisotropism and perfect cleavage. Colour varieties include recently discovered orange kyanite from Tanzania. The orange colour is due to inclusion of small amounts of manganese (Mn^{3+}) in the structure.

Kyanite is one of the index minerals that are used to estimate the temperature, depth, and pressure at which a rock undergoes metamorphism.

Notes for Identification

Kyanite's elongated, columnar crystals are usually a good first indication of the mineral, as well as its colour (when the specimen is blue). Associated minerals are useful as well, especially the presence of the polymorphs of staurolite, which occur frequently with kyanite. However, the most useful characteristic in identifying kyanite is its anisotropism. If one suspects a specimen to be kyanite, verifying that it has two distinctly different hardnesses on perpendicular axes is a key to identification.

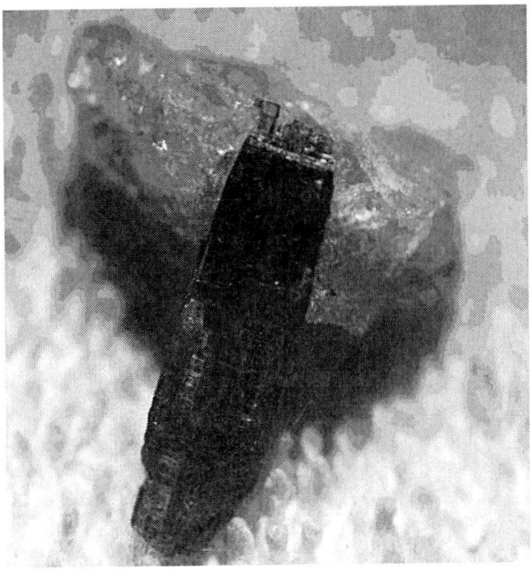

Figure 150: Deep blue kyanite

Occurrence

Kyanite occurs in gneiss, schist, pegmatite, and quartz veins resulting from moderate to high-pressure regional metamorphism of principally pelitic rocks. It occurs as detrital grains in sedimentary rocks. It occurs associated with staurolite, andalusite, sillimanite, talc, hornblende, gedrite, mullite and corundum.

Sillimanite

Figure 151: Sillimanite

Sillimanite is an alumino-silicate mineral with the chemical formula Al_2SiO_5. Sillimanite is named after the American chemist Benjamin Silliman (1779-1864). It was first described in 1824 for an occurrence in Chester, Middlesex County, Connecticut, USA.

Occurrence

Figure 152: Sillimanite cut

Sillimanite is one of three alumino-silicate polymorphs, the other two being andalusite and kyanite. A common variety of sillimanite is known as *fibrolite,* so named because the mineral appears like a bunch of fibres twisted together when viewed in thin section or even by the naked eye.

Both the fibrous and traditional forms of sillimanite are common in metamorphosed sedimentary rocks. It is an index mineral indicating high-temperature metamorphism. It occurs with andalusite, kyanite, potassium feldspar, almandine, cordierite, biotite and quartz in schist, gneiss, hornfels and also rarely in pegmatites.

Sillimanite has been found in Brandywine Springs, New Castle County, Delaware, USA. It was named by the State Legislature in 1977 as the state mineral of Delaware by suggestion of the Delaware Mineralogical Society, Inc.

Natural sillimanite rocks cut into the required shape and size are used mainly in glass industries. Sillimanite is the best raw material for the manufacture of high alumina refractories or 55-60 % alumina bricks. But its use on large scale is not possible due to its fine grading and high cost. Dumortierite and mullite are similar mineral species used in quality porcelain.

Limestone

Figure 153: Limestone in Waitomo District, New Zealand

Limestone is a sedimentary rock composed largely of the minerals calcite and aragonite, which are different crystal forms of calcium carbonate ($CaCO_3$). Many limestones are composed from skeletal fragments of marine organisms such as coral or foraminifera.

Limestone makes up about 10% of the total volume of all sedimentary rocks. The solubility of limestone in water and weak acid solutions leads to karst landscapes, in which water erodes the limestone over thousands to millions of years. Most cave systems are through limestone bedrock. Limestone has numerous uses, including as a building material, as aggregate to form the base of roads, as white pigment or filler in products such as toothpaste or paints, and as a chemical feedstock.

Description

Figure 154: Limestone quarry at Cedar Creek, Virginia, USA

Figure 155: La Zaplaz formations in the Piatra Craiului Mountains, Romania.

Some limestones do not consist of grains at all, and are formed completely by the chemical precipitation of calcite or aragonite, i.e. travertine. Secondary calcite may be deposited by supersaturated meteoric waters (groundwater that precipitates the material in caves). This produces speleothems, such as stalagmites and stalactites. Another form taken by calcite is oolitic limestone, which can be recognised by its granular (oolite) appearance.

The primary source of the calcite in limestone is most commonly marine organisms. Some of these organisms can construct mounds of rock known as reefs, building upon past generations. Below about 3,000 metres, water pressure and temperature conditions cause the dissolution of calcite to increase nonlinearly, so limestone typically does not form in deeper waters. Limestones may also form in both lacustrine and evaporite depositional environments.

Calcite can be either dissolved or precipitated by groundwater, depending on several factors, including the water temperature, pH, and dissolved ion concentrations. Calcite exhibits an unusual characteristic called retrograde solubility, in which it becomes less soluble in water as the temperature increases.

Because of impurities, such as clay, sand, organic remains, iron oxide and other materials, many limestones exhibit different colours, especially on weathered surfaces.

Limestone may be crystalline, clastic, granular, or massive, depending on the method of formation. Crystals of calcite, quartz, dolomite or barite may line small cavities in the rock. When conditions are right for precipitation, calcite forms mineral coatings that cement the existing rock grains together, or it can fill fractures.

Travertine is a banded, compact variety of limestone formed along streams, particularly where there are waterfalls, and around hot or cold springs. Calcium carbonate is deposited where evaporation of the water leaves a solution supersaturated with the chemical constituents of calcite. Tufa, a porous or cellular variety of travertine, is found near waterfalls. Coquina is a poorly consolidated limestone composed of pieces of coral or shells.

During regional metamorphism that occurs during the mountain building process (orogeny), limestone recrystallises into marble.

Limestone is a parent material of Mollisol soil group.

Classification

Two major classification schemes, the Folk and the Dunham, are used for identifying limestone and carbonate rocks.

Folk Classification

Robert L. Folk developed a classification system that places primary emphasis on the detailed composition of grains and interstitial material in carbonate rocks. Based on composition, there are three main components: allochems (grains), matrix (mostly micrite), and cement (sparite).

The Folk system uses two-part names; the first refers to the grains and the second is the root. It is helpful to have a petrographic microscope when using the Folk scheme, because it is easier to determine the components present in each sample.

Dunham Classification

The Dunham scheme focuses on depositional textures. Each name is based upon the texture of the grains that make up the limestone. Robert J. Dunham published his system for limestone in 1962; it focuses on the depositional fabric of carbonate rocks. Dunham divides the rocks into four main groups based on relative proportions of coarser clastic particles.

Dunham names are essentially for rock families. His efforts deal with the question of whether or not the grains were originally in mutual contact, and therefore self-supporting, or whether the rock is characterised by the presence of frame builders and algal mats.

Unlike the Folk scheme, Dunham deals with the original porosity of the rock. The Dunham scheme is more useful for hand samples because it is based on texture, not the grains in the sample.

Types

There are many types of limestone the most common ones: Chalk, crystalline, fossilferous, oolitic and travertine.

Limestone Landscape

Figure 156: *The Cudgel of Hercules*, a tall limestone rock (Pieskowa Ska³a Castle in the background).

Limestone makes up about 10% of the total volume of all sedimentary rocks.

Limestone is partially soluble, especially in acid, and therefore forms many erosional landforms. These include limestone pavements, pot holes, cenotes, caves and gorges.

Such erosion landscapes are known as karsts. Limestone is less resistant than most igneous rocks, but more resistant than most other sedimentary rocks. It is therefore usually associated with hills and downland, and occurs in regions with other sedimentary rocks, typically clays.

Karst topography and caves develop in limestone rocks due to their solubility in dilute acidic groundwater. The solubility of limestone in water and weak acid solutions leads to karst landscapes. Regions overlying limestone bedrock tend to have fewer visible above-ground sources (ponds and streams), as surface water easily drains downward through joints in the limestone.

While draining, water and organic acid from the soil slowly (over thousands or millions of years) enlarges these cracks, dissolving the calcium carbonate and carrying it away in solution. Most cave systems are through limestone bedrock. Cooling groundwater or mixing of different groundwaters will also create conditions suitable for cave formation.

Coastal limestones are often eroded by organisms which bore into the rock by various means. This process is known as bioerosion. It is most common in the tropics, and it is known throughout the fossil record.

Bands of limestone emerge from the Earth's surface in often spectacular rocky outcrops and islands. Examples include the Burren in Co.

Clare, Ireland; the Verdon Gorge in France; Malham Cove in North Yorkshire and the Isle of Wight, England; on Fårö near the Swedish island of Gotland, the Niagara Escarpment in Canada/United States, Notch Peak in Utah, the Ha Long Bay National Park in Vietnam and the hills around the Lijiang River and Guilin city in China.

The Florida Keys, islands off the south coast of Florida, are composed mainly of oolitic limestone (the Lower Keys) and the carbonate skeletons of coral reefs (the Upper Keys), which thrived in the area during interglacial periods when sea level was higher than at present.

Unique habitats are found on alvars, extremely level expanses of limestone with thin soil mantles. The largest such expanse in Europe is the Stora Alvaret on the island of Öland, Sweden. Another area with large quantities of limestone is the island of Gotland, Sweden. Huge quarries in northwestern Europe, such as those of Mount Saint Peter (Belgium/Netherlands), extend for more than a hundred kilometres.

The world's largest limestone quarry is at Michigan Limestone and Chemical Company in Rogers City, Michigan.

Uses

Limestone is very common in architecture, especially in Europe and North America. Many landmarks across the world, including the Great Pyramid and its associated complex in Giza, Egypt, are made of limestone. So many buildings in Kingston, Canada were constructed from it that it is nicknamed the 'Limestone City'.

Figure 157: The Great Pyramid of Giza, one of the Seven Wonders of the Ancient World; its outside cover is made entirely from limestone.

On the island of Malta, a variety of limestone called Globigerina limestone was, for a long time, the only building material available, and is still very frequently used on all types of buildings and sculptures. Limestone is readily available and relatively easy to cut into blocks or more elaborate carving.

Figure 158: Courthouse built of limestone in Manhattan, Kansas

It is also long-lasting and stands up well to exposure. However, it is a very heavy material, making it impractical for tall buildings, and relatively expensive as a building material.

Figure 159: A limestone plate with a negative map of Moosburg in Bavaria is prepared for a lithography print.

Limestone was most popular in the late 19th and early 20th centuries. Train stations, banks and other structures from that era are normally made of limestone. It is used as a facade on some skyscrapers, but only in thin plates for covering, rather than solid blocks. In the United States, Indiana, most notably the Bloomington area, has long been a source of high quality quarried limestone, called Indiana limestone. Many famous buildings in London are built from Portland limestone.

Limestone was also a very popular building block in the Middle Ages in the areas where it occurred, since it is hard, durable, and commonly occurs in easily accessible surface exposures. Many medieval churches and castles in Europe are made of limestone. Beer stone was a popular kind of limestone for medieval buildings in southern England.

Figure 160: Limestone cropping out at São Pedro de Moel beach, Marinha Grande, Portugal

Limestone and (to a lesser extent) marble are reactive to acid solutions, making acid rain a significant problem to the preservation of artifacts made from this stone. Many limestone statues and building surfaces have suffered severe damage due to acid rain. Acid-based cleaning chemicals can also etch limestone, which should only be cleaned with a neutral or mild alkaline-based cleaner.

Other uses include:

- It is the raw material for the manufacture of quicklime (calcium oxide), slaked lime (calcium hydroxide), cement and mortar.
- Pulverized limestone is used as a soil conditioner to neutralise acidic soils.
- It is crushed for use as aggregate—the solid base for many roads.

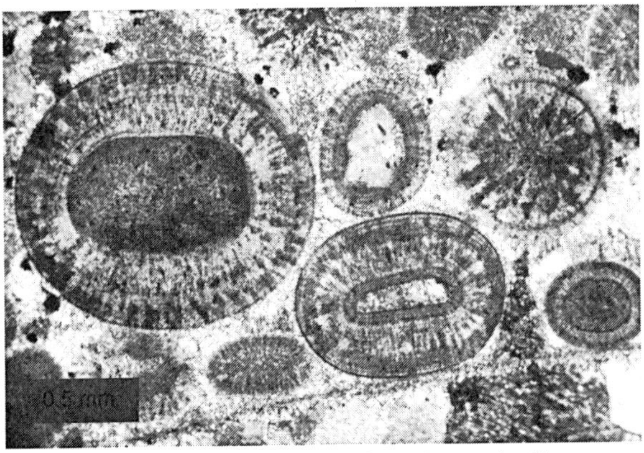

Figure 161: Thin-section view of a Middle Jurassic limestone in southern Utah. The round grains are ooids; the largest is 1.2 mm in diameter. This limestone is an oosparite.

Figure 162: Photo and etched section of a sample of fossiliferous limestone from the Kope Formation near Cincinnati, Ohio

- Geological formations of limestone are among the best petroleum reservoirs;
- As a reagent in flue gas desulfurization, it reacts with sulfur dioxide for air pollution control.
- Glass making, in some circumstances, uses limestone.
- It is added to toothpaste, paper, plastics, paint, tiles, and other materials as both white pigment and a cheap filler.
- It can suppress methane explosions in underground coal mines.
- Purified, it is added to bread and cereals as a source of calcium.
- Calcium levels in livestock feed are supplemented with it, such as for poultry (when ground up).
- It can be used for remineralising and increasing the alkalinity of purified water to prevent pipe corrosion and to restore essential nutrient levels.
- Used in blast furnaces, limestone extracts iron from its ore.
- It is often found in medicines and cosmetics.
- It is used in sculptures because of its suitability for carving.

Dolomite

Figure 163: Dolomite and magnesite – Spain

Dolomite is a carbonate mineral composed of calcium magnesium carbonate $CaMg(CO_3)_2$. The term is also used to describe the sedimentary carbonate rock dolostone.

Figure 164: Dolomite druse from Lawrence County, Arkansas, USA (size: 24×18×8 cm)

Figure 165: Dolomite.

Dolostone (dolomite rock) is composed predominantly of the mineral dolomite with a stoichiometric ratio of 50% or greater content of magnesium replacing calcium, often as a result of diagenesis. Limestone that is partially replaced by dolomite is referred to as *dolomitic limestone*, or in old U.S. geologic literature as *magnesian limestone*. Dolomite was first described in 1791 as a rock by the French naturalist and geologist, Déodat Gratet de Dolomieu (1750–1801) from exposures in what are now known as the Dolomite Alps of northern Italy.

Properties

The mineral dolomite crystallises in the trigonal-rhombohedral system. It forms white, gray to pink, commonly curved (saddle shape) crystals, although it is usually massive. Unlike calcite, dolomite is a double carbonate, having a different structural arrangement, and it does not rapidly dissolve or effervesce (fizz) in dilute hydrochloric acid unless it is scratched or in powdered form.

Crystal twinning is common. A solid solution series exists between dolomite and iron rich ankerite. Small amounts of iron in the structure give the crystals a yellow to brown tint. Manganese substitutes in the structure also up to about three percent MnO. A high manganese content gives the crystals a rosy pink colour noted in the image above. A series with the manganese rich kutnohorite may exist. Lead and zinc also substitute in the structure for magnesium.

Formation

Figure 166: Dolomite bedrock underneath a Bristlecone Pine, White Mountains, California.

Vast deposits are present in the geological record, but the mineral is relatively rare in modern environments. Laboratory synthesis of stoichiometric dolomite has been carried out only at temperatures of greater than 100 °C (conditions typical of burial in sedimentary basins), even though much dolomite in the rock record appears to have formed in low-temperature conditions.

The high temperature is likely to speed up the movement of calcium and magnesium ions so that they can find their places in the ordered structure within a reasonable amount of time.

This suggests that the lack of dolomite that is being formed today is likely due to kinetic factors, i.e. due to the lack of kinetic energy or temperature.

Modern dolomite does occur as a precipitating mineral in specialised environments on the surface of the earth today. In the 1950s and 60s, dolomite was found to be forming in highly saline lakes in the Coorong region of South Australia. Dolomite crystals also occur in deep-sea sediments, where organic matter content is high. This dolomite is termed "organogenic" dolomite.

Recent research has found modern dolomite formation under anaerobic conditions in supersaturated saline lagoons along the Rio de Janeiro coast of Brazil, namely, Lagoa Vermelha and Brejo do Espinho. One interesting reported case was the formation of dolomite in the kidneys of a Dalmatian dog.

This was believed to be due to chemical processes triggered by bacteria. Dolomite has been speculated to develop under these conditions with the help of sulfate-reducing bacteria (e.g. *Desulfovibrio brasiliensis*).

The actual role of bacteria in the low-temperature formation of dolomite remains to be demonstrated. The specific mechanism of dolomitization, involving sulfate-reducing bacteria, has not yet been demonstrated.

Dolomite appears to form in many different types of environment and can have varying structural, textural and chemical characteristics. Some researchers have stated "there are dolomites and dolomites", meaning that there may not be one single mechanism by which dolomite can form. Much modern dolomite differs significantly from the bulk of the dolomite found in the rock record, leading researchers to speculate that environments where dolomite formed in the geologic past differ significantly from those where it forms today.

Reproducible laboratory syntheses of dolomite (and magnesite) leads first to the initial precipitation of a metastable "precursor" (such as magnesium calcite), to be changed gradually into more and more of the stable phase (such as dolomite or magnesite) during periodical intervals of dissolution and reprecipitation. The general principle governing the course of this irreversible geochemical reaction has been coined Ostwald's step rule.

For a very long time scientists had difficulties synthesizing dolomite. However, in a 1999 study, through a process of dissolution alternating with intervals of precipitation, measurable levels of dolomite were synthesized at low temperatures and pressures.

Coral Atolls

Dolomitization of calcite also occurs at certain depths of coral atolls where water is undersaturated in calcium carbonate but saturated in dolomite. Convection created by tides and sea currents enhance this change. Hydrothermal currents created by volcanoes under the atoll may also play an important role.

Uses

Figure 167: Dolomite with chalcopyrite from the Tri-state district, Cherokee County, Kansas (size: 11.4×7.2×4.6 cm)

Dolomite is used as an ornamental stone, a concrete aggregate, a source of magnesium oxide and in the Pidgeon process for the production of magnesium.

It is an important petroleum reservoir rock, and serves as the host rock for large strata-bound Mississippi Valley-Type (MVT) ore deposits of base metals such as lead, zinc, and copper.

Where calcite limestone is uncommon or too costly, dolomite is sometimes used in its place as a flux for the smelting of iron and steel. Large quantities of processed dolomite are used in the production of float glass.

In horticulture, dolomite and dolomitic limestone are added to soils and soilless potting mixes to lower their acidity and as a magnesium source. Home and container gardening are common examples of this use.

Dolomite is also used as the substrate in marine (saltwater) aquariums to help buffer changes in pH of the water.

Because dolomite contains relatively minor quantities of radioactive materials, physicists in particle physics build their particle detectors under layers of dolomite to enable the detectors to detect the highest possible number of exotic particles. Another reason to build their particle detectors under layers of dolomite is for its insulation against cosmic rays.

Marble

Figure 168: Folded and weathered marble at General Carrera Lake, Chile.

Marble is a metamorphic rock composed of recrystallised carbonate minerals, most commonly calcite or dolomite.

Geologists use the term "marble" to refer to metamorphosed limestone; however stonemasons use the term more broadly to encompass unmetamorphosed limestone.

Marble is commonly used for sculpture and as a building material.

Etymology

Whilst the English term resembles the French *marbre*, most other European languages (e.g. Spanish *mármol*, Italian *marmo*, Portuguese *mármore*, German, Norwegian, Danish and Swedish *marmor*, Dutch *marmer*, Polish *marmur*, Turkish *mermer*, Czech *mramor* and Russian *ìðáìîð*) follow the original Greek.

Physical Origins

Marble is a rock resulting from metamorphism of sedimentary carbonate rocks, most commonly limestone or dolomite rock.

Metamorphism causes variable recrystallisation of the original carbonate mineral grains.

The resulting marble rock is typically composed of an interlocking mosaic of carbonate crystals. Primary sedimentary textures and structures of the original carbonate rock (protolith) have typically been modified or destroyed.

Figure 169: The Taj Mahal is made of marble.

Pure white marble is the result of metamorphism of a very pure (silicate-poor) limestone or dolomite protolith. The characteristic swirls and veins of many coloured marble varieties are usually due to various mineral impurities such as clay, silt, sand, iron oxides, or chert which were originally present as grains or layers in the limestone.

Green colouration is often due to serpentine resulting from originally high magnesium limestone or dolostone with silica impurities. These various impurities have been mobilised and recrystallised by the intense pressure and heat of the metamorphism.

Uses

Sculpture: White marble has been prized for its use in sculptures since classical times. This preference has to do with its softness, relative isotropy and homogeneity, and a relative resistance to shattering. Also, the low index of refraction of calcite allows light to penetrate several millimetres into the stone before being scattered out, resulting in the characteristic waxy look which gives "life" to marble sculptures of the human body.

Construction Marble

Construction marble is a stone which is composed of calcite, dolomite or serpentine which is capable of taking a polish. More generally in construction, specifically the dimension stone trade, the term "marble" is used for any crystalline calcitic rock (and some non-calcitic rocks) useful as building stone.

For example, Tennessee marble is really a dense granular fossiliferous gray to pink to maroon Ordovician limestone that geologists call the Holston Formation.

Production

According to the United States Geological Survey, U.S. dimension marble production in 2006 was 46,400 tons valued at $18.1 million, compared to 72,300 tons valued at $18.9 million in 2005.

Crushed marble production (for aggregate and industrial uses) in 2006 was 11.8 million tons valued at $116 million, of which 6.5 million tons was finely ground calcium carbonate and the rest was construction aggregate.

For comparison, 2005 crushed marble production was 7.76 million tons valued at $58.7 million, of which 4.8 million tons was finely ground calcium carbonate and the rest was construction aggregate. U.S. dimension marble demand is about 1.3 million tons.

The DSAN World Demand for (finished) Marble Index has shown a growth of 12% annually for the 2000–2006 period, compared to 10.5% annually for the 2000–2005 period.

The largest dimension marble application is tile.

Pakistan is one of the largest marble exporters of the world with exports totalling to around a 100,000 tonnes per annum.

China is the chief importer of marble, specifically Pakistani marble, with imports amounting to more than 70,000 tonnes in a single calendar year.

Artificial Marble

Marble dust is combined with cement or synthetic resins to make *reconstituted* or *cultured marble.*

The appearance of marble can be simulated with faux marbling, a painting technique that imitates the stone's colour patterns.

Cultural Associations

Figure 170: Marble from Italy.

Figure 171: Ancient marble columns in the prayer hall of the Mosque of Uqba, in Kairouan, Tunisia

Graphite

The mineral graphite is one of the allotropes of carbon. It was named by Abraham Gottlob Werner in 1789 from the Ancient Greek (*graphô*), "to draw/write", for its use in pencils, where it is commonly called lead (not to be confused with the metallic element lead). Unlike diamond (another carbon allotrope), graphite is an electrical conductor, a semimetal. It is, consequently, useful in such applications as arc lamp electrodes.

Graphite is the most stable form of carbon under standard conditions. Therefore, it is used in thermochemistry as the standard state for defining the heat of formation of carbon compounds. Graphite may be considered the highest grade of coal, just above anthracite and alternatively called meta-anthracite, although it is not normally used as fuel because it is difficult to ignite.

There are three principal types of natural graphite, each occurring in different types of ore deposit:

1. Crystalline flake graphite (or flake graphite for short) occurs as isolated, flat, plate-like particles with hexagonal edges if unbroken and when broken the edges can be irregular or angular;
2. Amorphous graphite occurs as fine particles and is the result of thermal metamorphism of coal, the last stage of coalification, and is sometimes called meta-anthracite. Very fine flake graphite is sometimes called amorphous in the trade;
3. Lump graphite (also called vein graphite) occurs in fissure veins or fractures and appears as massive platy intergrowths of fibrous or acicular crystalline aggregates, and is probably hydrothermal in origin.

Highly ordered pyrolytic graphite or highly oriented pyrolytic graphite (HOPG) refers to graphite with an angular spread between the graphite sheets of less than 1°. This highest-quality synthetic form is used in scientific research. The name "graphite fibre" is also sometimes used to refer to carbon fibre or carbon fibre-reinforced polymer.

Occurrence

Graphite occurs in metamorphic rocks as a result of the reduction of sedimentary carbon compounds during metamorphism. It also occurs in igneous rocks and in meteorites. Minerals associated with graphite include quartz, calcite, micas and tourmaline. In meteorites it occurs with troilite and silicate minerals.

According to the United States Geological Survey (USGS), world production of natural graphite in 2008 was 1,110 thousand tonnes (kt), of which the following major exporters are: China (800 kt), India (130 kt), Brazil (76 kt), North Korea (30 kt) and Canada (28 kt). Graphite is not mined in US, but US production of synthetic graphite in 2007 was 198 kt valued at $1.18 billion. US graphite consumption was 42 kt and 200 kt for natural and synthetic graphite, respectively.

Properties

Structure: Graphite has a layered, planar structure. In each layer, the carbon atoms are arranged in a hexagonal lattice with separation of 0.142 nm, and the distance between planes is 0.335 nm. The two known forms of graphite, *alpha* (hexagonal) and *beta* (rhombohedral), have very similar physical properties (except that the graphene layers stack slightly differently). The hexagonal graphite may be either flat or buckled. The alpha form can be converted to the beta form through mechanical treatment and the beta form reverts to the alpha form when it is heated above 1300 °C. The layering contributes to its lower density.

Other Properties

The acoustic and thermal properties of graphite are highly anisotropic, since phonons propagate very quickly along the tightly-bound planes, but are slower to travel from one plane to another.

Graphite can conduct electricity due to the vast electron delocalisation within the carbon layers (a phenomenon called aromaticity). These valence electrons are free to move, so are able to conduct electricity. However, the electricity is primarily conducted within the plane of the layers. The conductive properties of powdered graphite allowed its use as a semiconductor substitute in early Carbon microphones.

Graphite and graphite powder are valued in industrial applications for its self-lubricating and dry lubricating properties. There is a common belief that graphite's lubricating properties are solely due to the loose interlamellar coupling between sheets in the structure. However, it has been shown that in a vacuum environment (such as in technologies for use in space), graphite is a very poor lubricant. This observation led to the discovery that the lubrication is due to the presence of fluids between the layers, such as air and water, which are naturally adsorbed from the environment.

This molecular property is unlike other layered, dry lubricants such as molybdenum disulfide. Recent studies suggest that an effect called superlubricity can also account for graphite's lubricating properties. The use of graphite is limited by its tendency to facilitate pitting corrosion in some stainless steel, and to promote galvanic corrosion between dissimilar metals (due to its electrical conductivity). It is also corrosive to aluminium in the presence of moisture.

For this reason, the US Air Force banned its use as a lubricant in aluminium aircraft, and discouraged its use in aluminium-containing automatic weapons.

Even graphite pencil marks on aluminium parts may facilitate corrosion. Another high-temperature lubricant, hexagonal boron nitride, has the same molecular structure as graphite. It is sometimes called *white graphite*, due to its similar properties.

When a large number of crystallographic defects bind these planes together, graphite loses its lubrication properties and becomes what is known as pyrolytic carbon. This material is useful for blood-contacting implants such as artificial heart valves. It is also highly diamagnetic, thus it will float in mid-air above a strong magnet.

Natural and crystalline graphites are not often used in pure form as structural materials, due to their shear-planes, brittleness and inconsistent mechanical properties.

History of Natural Graphite Use

Graphite was used by the 4th millennium B.C. Marıca culture to create a ceramic paint to decorate pottery during the Neolithic Age in southeastern Europe.

Some time before 1565 (some sources say as early as 1500), an enormous deposit of graphite was discovered on the approach to Grey Knotts from the hamlet of Seathwaite in Borrowdale parish, Cumbria, England, which the locals found very useful for marking sheep.

During the reign of Elizabeth 1 (1533-1603), Borrowdale graphite was used as a refractory material to line molds for cannon balls, resulting in rounder, smoother balls that could be fired further, contributing to the growing superiority of the English navy. This particular deposit of graphite was extremely pure and soft, and could easily be broken into sticks. Because of its military importance, this unique mine and its production were strictly controlled by the Crown.

Other Names

Historically, graphite was called blacklead and plumbago.

Plumbago was commonly used for its massive mineral form. Both of these names arise from confusion with the similar-appearing lead ores, particularly galena. The Latin word for lead is *plumbum*, which gave its name to both the English term for this grey metallic-sheened mineral and even the leadworts or plumbagos, plants with flowers that resemble this colour.

The term blacklead has usually been applied to a powdered or processed form, where this fine powder then appears as a matte non-metallic black.

Uses of Natural Graphite

Natural graphite is mostly consumed for refractories, steelmaking, expanded graphite, brake linings, foundry facings and lubricants. Graphene, which occurs naturally in graphite, has unique physical properties and might be one of the strongest substances known; however, the process of separating it from graphite will require some technological development before it is economically feasible to use it in industrial processes.

Refractories

This end-use begins before 1900 with the graphite crucible used to hold molten metal; this is now a minor part of refractories. In the mid 1980s, the carbon-magnesite brick became important, and a bit later the alumina-graphite shape. Currently the order of importance is alumina-graphite shapes, carbon-magnesite brick, monolithics (gunning and ramming mixes), and then crucibles.

Crucibles began using very large flake graphite, and carbon-magnesite brick requiring not quite so large flake graphite; for these and others there is now much more flexibility in size of flake required, and amorphous graphite is no longer restricted to low-end refractories. Alumina-graphite shapes are used as continuous casting ware, such as nozzles and troughs, to convey the molten steel from ladle to mold, and carbon magnesite bricks line steel converters and electric arc furnaces to withstand extreme temperatures. Graphite Blocks are also used in parts of blast furnace linings where the high thermal conductivity of the graphite is critical. High-purity monolithics are often used as a continuous furnace lining instead of the carbon-magnesite bricks.

The US and European refractories industry had a crisis in 2000–2003, with an indifferent market for steel and a declining refractory consumption per tonne of steel underlying firm buyouts and many plant closings.

Many of the plant closings resulted from the acquisition of Harbison-Walker Refractories by RHI AG some plants had their equipment auctioned off.

Since much of the lost capacity was for carbon-magnesite brick, graphite consumption within refractories area moved towards alumina-graphite shapes and monolithics, and away from the brick. The major source of carbon-magnesite brick is now imports from China. Almost all of the above refractories are used to make steel and account for 75% of refractory consumption; the rest is used by a variety of industries, such as cement.

According to the USGS, US natural graphite consumption in refractories was 11,000 tonnes in 2006.

Steelmaking

Natural graphite in this end use mostly goes into carbon raising in molten steel, although it can be used to lubricate the dies used to extrude hot steel. Supplying carbon raisers is very competitive, therefore subject to cut-throat pricing from alternatives such as synthetic graphite powder, petroleum coke, and other forms of carbon. A carbon raiser is added to increase the carbon content of the steel to the specified level. An estimate based on USGS US graphite consumption statistics indicates that 10,500 tonnes were used in this fashion in 2005.

Expanded Graphite

Expanded graphite is made by immersing natural flake graphite in a bath of chromic acid, then concentrated sulfuric acid, which forces the crystal lattice planes apart, thus expanding the graphite. The expanded graphite can be used to make graphite foil or used directly as "hot top" compound to insulate molten metal in a ladle or red-hot steel ingots and decrease heat loss, or as firestops fitted around a fire door or in sheet metal collars surrounding plastic pipe (during a fire, the graphite expands and chars to resist fire penetration and spread), or to make high-performance gasket material for high-temperature use. After being made into graphite foil, the foil is machined and assembled into the bipolar plates in fuel cells.

The foil is made into heat sinks for laptop computers which keeps them cool while saving weight, and is made into a foil laminate that can be used in valve packings or made into gaskets. Old-style packings are now a minor member of this grouping: fine flake graphite in oils or greases for uses requiring heat resistance. A GAN estimate of current US natural graphite consumption in this end use is 7,500 tonnes.

Intercalated Graphite

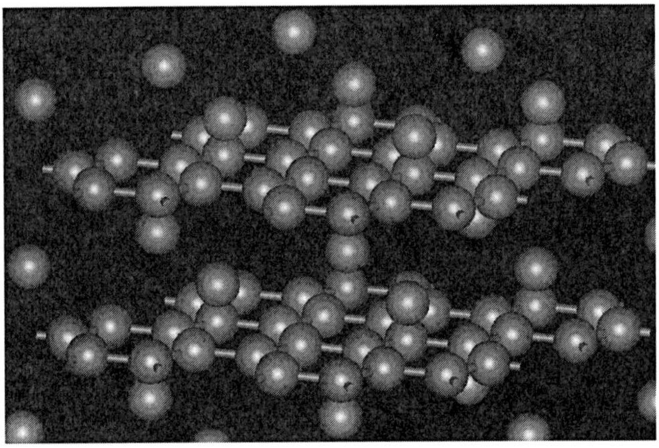

Figure 172: Structure of CaC_6

Graphite forms intercalation compounds with some metals and small molecules. In these compounds, the host molecule or atom gets "sandwiched" between the graphite layers, resulting in compounds with variable stoichiometry. A prominent example of an intercalation compound is potassium graphite, denoted by the formula KC_8. Graphite intercalation compounds are superconductors. The highest transition temperature (by June 2009) T_c = 11.5 K is achieved in CaC_6 and it further increases under applied pressure (15.1 K at 8 GPa).

Brake Linings

Natural amorphous and fine flake graphite are used in brake linings or brake shoes for heavier (nonautomotive) vehicles, and became important with the need to substitute for asbestos. This use has been important for quite some time, but nonasbestos organic (NAO) compositions are beginning to cost graphite market share. A brake-lining industry shake-out with some plant closings has not helped either, nor has an indifferent automotive market. According to the USGS, US natural graphite consumption in brake linings was 6,510 tonnes in 2005.

Foundry Facings and Lubricants

A foundry facing mold wash is a water-based paint of amorphous or fine flake graphite. Painting the inside of a mold with it and letting it dry leaves a fine graphite coat that will ease separation of the object cast after the hot metal has cooled.

Graphite lubricants are speciality items for use at very high or very low temperatures, as forging die lubricant, an antiseize agent, a gear lubricant for mining machinery, and to lubricate locks. Having low-grit graphite, or even better no-grit graphite (ultra high purity), is highly desirable.

It can be used as a dry powder, in water or oil, or as colloidal graphite (a permanent suspension in a liquid). An estimate based on USGS graphite consumption statistics indicates that 2,200 tonnes was used in this fashion in 2005.

Other Uses

Natural graphite has found uses as the marking material ("lead") in common pencils, in zinc-carbon batteries, in electric motor brushes, and various specialised applications.

Uses of Synthetic Graphite

Invention of Synthetic Graphite: Synthetic graphite was invented/discovered by Edward Goodrich Acheson (1856 - 1931). In the mid 1890s, Acheson discovered that overheating carborundum (also called silicon carbide or SiC, which he also invented/discovered) produced almost pure graphite.

While studying the effects of high temperature on carborundum, he had found that silicon vaporizes at about 4,150° C (7,500° F), leaving behind graphitic carbon. This graphite was another major discovery for him, and it became extremely valuable and helpful as a lubricant. The Acheson Graphite Co. was formed in 1899. In 1928 this company was merged with National Carbon Co (now Union Carbide). Acheson also developed a variety of colloidal graphite products including Oildag and Aquadag. These were later manufactured by the Acheson Colloids Co. (now Acheson Industries).

Electrodes

These electrodes carry the electricity that melts scrap iron and steel (and sometimes direct-reduced iron: DRI) in electric arc furnaces, the vast majority of steel furnaces. They are made from petroleum coke after it is mixed with coal tar pitch, extruded and shaped, then baked to carbonize the binder (pitch), and then graphitized by heating it to temperatures approaching 3000°C, that converts carbon to graphite. They can vary in size up to 11 ft. long and 30 in. in diametre.

An increasing proportion of global steel is made using electric arc furnaces, and the electric arc furnace itself is getting more efficient and making more steel per tonne of electrode. An estimate based on USGS data indicates that graphite electrode consumption was 197,000 tonnes in 2005.

Powder and Scrap

The powder is made by heating powdered petroleum coke above the temperature of graphitization, sometimes with minor modifications. The graphite scrap comes from pieces of unusable electrode material (in the manufacturing stage or after use) and lathe turnings, usually after crushing and sizing.

Most synthetic graphite powder goes to carbon raising in steel (competing with natural graphite), with some used in batteries and brake linings. According to the USGS, US synthetic graphite powder and scrap production was 95,000 tonnes in 2001 (latest data).

Neutron Moderator

Special grades of synthetic graphite also find use as a matrix and neutron moderator within nuclear reactors. Its low neutron cross-section also recommends it for use in proposed fusion reactors. Care must be taken that reactor-grade graphite is free of neutron absorbing materials such as boron, widely used as the seed electrode in commercial graphite deposition systems—this caused the failure of the Germans' World War II graphite-based nuclear reactors.

Since they could not isolate the difficulty they were forced to use far more expensive heavy water moderators. Graphite used for nuclear reactors is often referred to as nuclear graphite.

Other Uses

Graphite (carbon) fibre and carbon nanotubes are also used in carbon fibre reinforced plastics, and in heat-resistant composites such as reinforced carbon-carbon (RCC). Commercial structures made from carbon fibre graphite composites include fishing rods, golf club shafts, bicycle frames, sports car body panels, the fuselage of the Boeing 787 Dreamliner and pool cue sticks and have been successfully employed in reinforced concrete, The mechanical properties of carbon fibre graphite-reinforced plastic composites and grey cast iron are strongly influenced by the role of graphite in these materials.

In this context, the term "(100%) graphite" is often loosely used to refer to a pure mixture of carbon reinforcement and resin, while the term "composite" is used for composite materials with additional ingredients.

Graphite has been used in at least three radar absorbent materials. It was mixed with rubber in Sumpf and Schornsteinfeger, which were used on U-boat snorkels to reduce their radar cross section. It was also used in tiles on early F-117 Nighthawks. Modern smokeless powder is coated in graphite to prevent the buildup of static charge.

Graphite Mining, Beneficiation, and Milling

Graphite is mined around the world by both open pit and underground methods. While flake graphite and amorphous graphite are both mined open pit and underground, lump (vein) graphite is only mined underground in Sri Lanka.

The open pit mines usually employ equipment (i.e. bulldozers) to scoop up the ore, which is usually put in trucks and moved to the plant. Since the original rock is usually lateritized or weathered, this amounts to moving dirt with flecks or pieces of graphite in it from the pit (blasting is seldom required). The underground graphite mines employ drilling and blasting to break up the hard rock (ore), which is then moved by mine cars pulled by a locomotive, or moved by automotive vehicles, to the surface and then to the plant. In less-developed areas of the world, the ore can be mined by pick and shovel and transported by mine cars pushed by a labourer or by women carrying baskets of ore on their heads.

Graphite usually needs beneficiation, although thick-bedded amorphous graphite and vein graphite is almost always beneficiated, if beneficiated at all, by labourers hand-picking out the pieces of gangue (rock) and hand-screening the product. The great majority of world flake graphite production is crushed and ground if necessary and beneficiated by flotation. Treating graphite by flotation encounters one big difficulty: graphite is very soft and “marks” (coats) the particles of gangue. This makes the “marked” gangue particles float off with the graphite to yield a very impure concentrate.

There are two ways of obtaining a saleable concentrate or product: regrinding and floating it again and again (up to seven times) to obtain a purer and purer concentrate, or by leaching (dissolving) the gangue with hydrofluoric acid (for a silicate gangue) or hydrochloric acid (for a carbonate gangue).

In the milling process, the incoming graphite products and concentrates can be ground before being classified (sized or screened), with the coarser flake size fractions (below 8 mesh, 8–20 mesh, 20–50 mesh) carefully preserved, and then the carbon contents are determined. Then some standard blends can be prepared from the different fractions, each with a certain flake size distribution and carbon content.

Custom blends can also be made for individual customers who want a certain flake size distribution and carbon content. If flake size is unimportant, the concentrate can be ground more freely. Typical final products include a fine powder for use as a slurry in oil drilling; in zirconium silicate, sodium silicate and isopropyl alcohol coatings for foundry molds; and a carbon raiser in the steel industry (Synthetic graphite powder and powdered petroleum coke can also be used as carbon raiser).

Rough graphite is typically classified, ground, and packaged at a graphite mill; often the more complex formulations are also mixed and packaged at the mill facility. Environmental impacts from graphite mills consist of air pollution including fine particulate exposure of workers and also soil contamination from powder spillages leading to heavy metals contaminations of soil. Dust masks are normally worn by workers during the production process to avoid worker exposure to the fine airborne graphite and zircon silicate.

Graphite Recycling

The most common way graphite is recycled occurs when synthetic graphite electrodes are either manufactured and pieces are cut off or lathe turnings are discarded, or the electrode (or other) are used all the way down to the electrode holder. A new electrode replaces the old one, but a sizeable piece of the old electrode remains. This is crushed and sized, and the resulting graphite powder is mostly used to raise the carbon content of molten steel. Graphite-containing refractories are sometimes also recycled, but often not because of their graphite: the largest-volume items, such as carbon-magnesite bricks that contain only 15–25% graphite, usually contain too little graphite. However, some recycled carbon-magnesite brick is used as the basis for furnace repair materials, and also crushed carbon-magnesite brick is used in slag conditioners. While crucibles have a high graphite content, the volume of crucibles used and then recycled is very small.

A high-quality flake graphite product that closely resembles natural flake graphite can be made from steelmaking kish. Kish is a large-volume near-molten waste skimmed from the molten iron feed to a basic oxygen furnace, and is a mix of graphite (precipitated out of the supersaturated iron), lime-rich slag, and some iron. The iron is recycled on site, so what is left is a mixture of graphite and slag. The best recovery process uses hydraulic classification (Which utilises a flow of water to separate minerals by specific gravity: graphite is light and settles nearly last.) to get a 70% graphite rough concentrate. Leaching this concentrate with hydrochloric acid gives a 95% graphite product with a flake size ranging from 10 mesh down.

Chapter 3

Exploration of Minerals

Petroleum or crude oil is a naturally occurring, flammable liquid consisting of a complex mixture of hydrocarbons of various molecular weights and other liquid organic compounds, that are found in geologic formations beneath the Earth's surface. Petroleum is recovered mostly through oil drilling. This latter stage comes after the studies of structural geology (at the reservoir scale), sedimentary basin analysis, reservoir characterization (mainly in terms of porosity and permeable structures). It is refined and separated, most easily by boiling point, into a large number of consumer products, from petrol and kerosene to asphalt and chemical reagents used to make plastics and pharmaceuticals. Petroleum is used in manufacturing a wide variety of materials.

The term *petroleum* was found (in the spelling "petraoleum") in 10th-century Old English sources. It was used in the treatise *De Natura Fossilium*, published in 1546 by the German mineralogist Georg Bauer, also known as Georgius Agricola. In the 19th century, the term *petroleum* was frequently used to refer to mineral oils produced by distillation from mined organic solids such as cannel coal (and later oil shale), and refined oils produced from them; in the United Kingdom, storage (and later transport) of these oils were regulated by a series of Petroleum Acts, from the *Petroleum Act 1862* c. 66 onward.

Composition

In its strictest sense, petroleum includes only crude oil, but in common usage it includes all liquid, gaseous, and solid (e.g., paraffin) hydrocarbons. Under surface pressure and temperature conditions, lighter hydrocarbons methane, ethane, propane and butane occur as gases, while pentane and heavier ones are in the form of liquids or solids. However, in an underground oil reservoir the proportions of gas, liquid, and solid depend on subsurface conditions and on the phase diagram of the petroleum mixture.

An oil well produces predominantly crude oil, with some natural gas dissolved in it. Because the pressure is lower at the surface than underground, some of the gas will come out of solution and be recovered (or burned) as *associated gas* or *solution gas*. A gas well produces predominantly natural gas.

However, because the underground temperature and pressure are higher than at the surface, the gas may contain heavier hydrocarbons such as pentane, hexane, and heptane in the gaseous state. At surface conditions these will condense out of the gas to form natural gas condensate, often shortened to *condensate*. Condensate resembles petrol in appearance and is similar in composition to some volatile light crude oils.

The proportion of light hydrocarbons in the petroleum mixture varies greatly among different oil fields, ranging from as much as 97% by weight in the lighter oils to as little as 50% in the heavier oils and bitumens.

The hydrocarbons in crude oil are mostly alkanes, cycloalkanes and various aromatic hydrocarbons while the other organic compounds contain nitrogen, oxygen and sulfur, and trace amounts of metals such as iron, nickel, copper and vanadium.

The exact molecular composition varies widely from formation to formation but the proportion of chemical elements vary over fairly narrow limits as follows:

Table 1: Composition by weight

Element	*Percent range*
Carbon	83 to 87%
Hydrogen	10 to 14%
Nitrogen	0.1 to 2%
Oxygen	0.05 to 1.5%
Sulfur	0.05 to 6.0%
Metals	< 0.1%

Four different types of hydrocarbon molecules appear in crude oil. The relative percentage of each varies from oil to oil, determining the properties of each oil.

Table 2: Composition by weight

Hydrocarbon	*Average*	*Range*
Paraffins	30%	15 to 60%
Naphthenes	49%	30 to 60%
Aromatics	15%	3 to 30%
Asphaltics	6%	remainder

Crude oil varies greatly in appearance depending on its composition. It is usually black or dark brown (although it may be yellowish, reddish, or even greenish). In the reservoir it is usually found in association with natural gas, which being lighter forms a gas cap over the petroleum, and saline water which, being heavier than most forms of crude oil, generally sinks beneath it.

Crude oil may also be found in semi-solid form mixed with sand and water, as in the Athabasca oil sands in Canada, where it is usually referred to as crude bitumen. In Canada, bitumen is considered a sticky, black, tar-like form of crude oil which is so thick and heavy that it must be heated or diluted before it will flow. Venezuela also has large amounts of oil in the Orinoco oil sands, although the hydrocarbons trapped in them are more fluid than in Canada and are usually called extra heavy oil.

These oil sands resources are called unconventional oil to distinguish them from oil which can be extracted using traditional oil well methods. Between them, Canada and Venezuela contain an estimated 3.6 trillion barrels (570×10^9 m^3) of bitumen and extra-heavy oil, about twice the volume of the world's reserves of conventional oil.

Petroleum is used mostly, by volume, for producing fuel oil and petrol, both important *"primary energy"* sources. 84% by volume of the hydrocarbons present in petroleum is converted into energy-rich fuels (petroleum-based fuels), including petrol, diesel, jet, heating, and other fuel oils, and liquefied petroleum gas.

The lighter grades of crude oil produce the best yields of these products, but as the world's reserves of light and medium oil are depleted, oil refineries are increasingly having to process heavy oil and bitumen, and use more complex and expensive methods to produce the products required. Because heavier crude oils have too much carbon and not enough hydrogen, these processes generally involve removing carbon from or adding hydrogen to the molecules, and using fluid catalytic cracking to convert the longer, more complex molecules in the oil to the shorter, simpler ones in the fuels.

Due to its high energy density, easy transportability and relative abundance, oil has become the world's most important source of energy since the mid-1950s. Petroleum is also the raw material for many chemical products, including pharmaceuticals, solvents, fertilizers, pesticides, and plastics; the 16% not used for energy production is converted into these other materials. Petroleum is found in porous rock formations in the upper strata of some areas of the Earth's crust.

There is also petroleum in oil sands (tar sands). Known oil reserves are typically estimated at around 190 km^3 (1.2 trillion (short scale) barrels) without oil sands, or 595 km^3 (3.74 trillion barrels) with oil sands. Consumption is currently around 84 million barrels (13.4×10^6 m^3) per day, or 4.9 km^3 per year. Which in turn yields a remaining oil supply of only about 120 years, if current demand remain static.

Chemistry

Petroleum is a mixture of a very large number of different hydrocarbons; the most commonly found molecules are alkanes (linear or branched), cycloalkanes, aromatic hydrocarbons, or more complicated chemicals like asphaltenes. Each petroleum variety has a unique mix of molecules, which define its physical and chemical properties, like colour and viscosity.

The *alkanes*, also known as *paraffins*, are saturated hydrocarbons with straight or branched chains which contain only carbon and hydrogen and have the general formula C_nH_{2n+2}. They generally have from 5 to 40 carbon atoms per molecule, although trace amounts of shorter or longer molecules may be present in the mixture.

The alkanes from pentane (C_5H_{12}) to octane (C_8H_{18}) are refined into petrol, the ones from nonane (C_9H_{20}) to hexadecane ($C_{16}H_{34}$) into diesel fuel, kerosene and jet fuel). Alkanes with more than 16 carbon atoms can be refined into fuel oil and lubricating oil. At the heavier end of the range, paraffin wax is an alkane with approximately 25 carbon atoms, while asphalt has 35 and up, although these are usually cracked by modern refineries into more valuable products. The shortest molecules, those with four or fewer carbon atoms, are in a gaseous state at room temperature.

They are the petroleum gases. Depending on demand and the cost of recovery, these gases are either flared off, sold as liquified petroleum gas under pressure, or used to power the refinery's own burners. During the winter, Butane (C_4H_{10}), is blended into the petrol pool at high rates, because butane's high vapour pressure assists with cold starts. Liquified under pressure slightly above atmospheric, it is best known for powering cigarette lighters, but it is also a main fuel source for many developing countries.

Propane can be liquified under modest pressure, and is consumed for just about every application relying on petroleum for energy, from cooking to heating to transportation.

The *cycloalkanes*, also known as *naphthenes*, are saturated hydrocarbons which have one or more carbon rings to which hydrogen atoms are attached according to the formula C_nH_{2n}. Cycloalkanes have similar properties to alkanes but have higher boiling points.

The *aromatic hydrocarbons* are unsaturated hydrocarbons which have one or more planar six-carbon rings called benzene rings, to which hydrogen atoms are attached with the formula C_nH_n. They tend to burn with a sooty flame, and many have a sweet aroma. Some are carcinogenic.

These different molecules are separated by fractional distillation at an oil refinery to produce petrol, jet fuel, kerosene, and other hydrocarbons. For example, 2,2,4-Trimethylpentane (isooctane), widely used in petrol, has a chemical formula of C_8H_{18} and it reacts with oxygen exothermically:

$$2\ C_8H_{18(l)} + 25\ O_{2(g)} \rightarrow 16\ CO_{2(g)} + 18\ H_2O_{(g)} + 10.86\ \text{MJ/mol (of octane)}$$

The amount of various molecules in an oil sample can be determined in laboratory. The molecules are typically extracted in a solvent, then separated in a gas chromatograph, and finally determined with a suitable detector, such as a flame ionization detector or a mass spectrometer. Due to the large number of co-eluted hydrocarbons within oil, many cannot be resolved by traditional gas chromatography and typically appear as a hump in the chromatogram. This unresolved complex mixture (UCM) of hydrocarbons is particularly apparent when analysing weathered oils and extracts from tissues of organisms exposed to oil.

Incomplete combustion of petroleum or petrol results in production of toxic byproducts. Too little oxygen results in carbon monoxide. Due to the high temperatures and high pressures involved, exhaust gases from petrol combustion in car engines usually include nitrogen oxides which are responsible for creation of photochemical smog.

Empirical Equations for the Thermal Properties of Petroleum Products

Heat of Combustion

At a constant volume the heat of combustion of a petroleum product can be approximated as follows:

$$Q_v = 12{,}400 - 2{,}100d^2$$

where Q_v is measured in cal/gram and d is the specific gravity at 60 °F (16 °C).

Thermal Conductivity

The thermal conductivity of petroleum based liquids can be modelled as follows: $K = \frac{0.813}{d}[1 - 0.0203(t - 32)]$

0.547

where K is measured in BTU · $hr^{-1}ft^{-2}$, t is measured in °F and d is the specific gravity at 60 °F (16 °C).

Specific Heat

The specific heat of a petroleum oils can be modelled as follows:

$$c = \frac{1}{\sqrt{d}}[0.388 + 0.00045t],$$

where c is measured in BTU/lbm-°F, t is the temperature in Fahrenheit and *d* is the specific gravity at 60 °F (16 °C).

In units of kcal/(kg·°C), the formula is:

$$L = \frac{1}{d}[110.9 - 0.09t],$$

where the temperature *t* is in Celsius and *d* is the specific gravity at 15 °C.

Latent Heat of Vaporization

The latent heat of vaporization can be modelled under atmospheric conditions as follows:

$$L = \frac{1}{d}\left[110.9 - 0.0t\right],$$

where L is measured in BTU/lbm, t is measured in °F and d is the specific gravity at 60 °F (16 °C).

In units of kcal/kg, the formula is:

$$L = \frac{1}{d}[194.4 - 0.162t],$$

where the temperature t is in Celsius and d is the specific gravity at 15 °C.

Formation

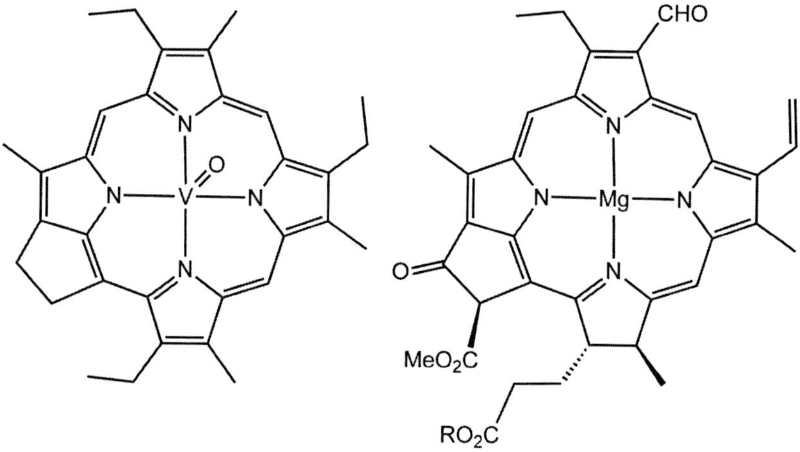

Figure 1: Structure of vanadium porphyrin compound extracted from petroleum by Alfred E. Treibs, father of organic geochemistry. Treibs noted the close structural similarity of this molecule and chlorophyll a.

Petroleum is a fossil fuel derived from ancient fossilized organic materials, such as zooplankton and algae. Vast quantities of these remains settled to sea or lake bottoms, mixing with sediments and being buried under anoxic conditions. As further layers settled to the sea or lake bed, intense heat and pressure built up in the lower regions.

This process caused the organic matter to change, first into a waxy material known as kerogen, which is found in various oil shales around the world, and then with more heat into liquid and gaseous hydrocarbons via a process known as catagenesis. Formation of petroleum occurs from hydrocarbon pyrolysis in a variety of mainly endothermic reactions at high temperature and/or pressure.

There were certain warm nutrient-rich environments such as the Gulf of Mexico and the ancient Tethys Sea where the large amounts of organic material falling to the ocean floor exceeded the rate at which it could decompose. This resulted in large masses of organic material being buried under subsequent deposits such as shale formed from mud. This massive organic deposit later became heated and transformed under pressure into oil.

Geologists often refer to the temperature range in which oil forms as an "oil window"—below the minimum temperature oil remains trapped in the form of kerogen, and above the maximum temperature the oil is converted to natural gas through the process of thermal cracking. Sometimes, oil formed at extreme depths may migrate and become trapped at a much shallower level. The Athabasca Oil Sands is one example of this.

Abiogenic Origin Hypothesis

Abiogenic petroleum origin is a largely abandoned hypothesis that was proposed as an alternative to theory of biological petroleum origin. It was relatively popular in the past, but it went largely forgotten at the end of the 20th century after it failed to predict the location of new wells.

The abiogenic hypothesis argues that petroleum was formed from deep carbon deposits, perhaps dating to the formation of the Earth. Supporters of the abiogenic hypothesis suggest that a great deal more petroleum exists on Earth than commonly thought, and that petroleum may originate from carbon-bearing fluids that migrate upward from the mantle. The presence of methane on Saturn's moon Titan and in the atmospheres of Jupiter, Saturn, Uranus and Neptune is cited as evidence of the formation of hydrocarbons without biology.

The biogenic theory for petroleum was first proposed by Georg Agricola in the 16th century and various abiogenic hypotheses were proposed in the 19th century, most notably by Alexander von Humboldt, the Russian chemist Dmitri Mendeleev and the French chemist Marcellin Berthelot. Abiogenic hypotheses were revived in the last half of the 20th century by Russian and Ukrainian scientists, who had little influence outside the Soviet Union because most of their research was published in their native languages. The theory was re-defined and made popular in the West by Thomas Gold, who published all his research in English.

Although the abiogenic hypothesis was accepted by many geologists in the former Soviet Union, it allegedly fell out of favour because it never made any useful prediction for the discovery of oil deposits. Most geologists now consider the abiogenic formation of petroleum scientifically unsupported. The abiogenic origin of petroleum has also recently been reviewed in detail by Glasby, who raises a number of objections, including that there is no direct evidence to date of abiogenic *petroleum* (liquid crude oil and long-chain hydrocarbon compounds).

It has been recently discovered that thermophilic bacteria, in the sea bottom and in cooling magma, produce methane and hydrocarbon gases, but studies indicate they are not produced in commercially significant quantities (i.e. in extracted hydrocarbon gases, the median abiogenic hydrocarbon content is 0.02%). A small number of geologists adhere to the abiogenic petroleum origin hypothesis, maintaining that high molecular weight hydrocarbons of purely inorganic origin exist within Earth's interior and are the source for major petroleum deposits. The hypothesis was originally proposed by Nikolai Kudryavtsev and Vladimir Porfiriev in the 1950s, and more recently Thomas Gold proposed a similar *deep hot biosphere* idea. The thermodynamic synthesis routes necessary to carry abiogenic source material into subsurface oil are not established, observation of organic markers in kerogen and oil is not explained, and no oil deposits have been located by this hypothesis.

History of Abiogenic Hypothesis

The abiogenic theory for the origin of petroleum is usually traced to the early part of the 19th century. The hypothesis was developed well before the field of organic chemistry, much less that of biochemistry, was established so the chemical nature of the petroleum was not known. Absent intellectual framework of organic and biological chemistry, abiologic theories were inevitable. In the early 19th century, Phlogiston theory was the dominant model for explaining chemical phenomena. Furthermore, the formal study of paleontology had only started in the early 19th century. It is within this scientifically primitive but changing environment that theories on the origin of petroleum originated.

Alexander von Humboldt was the first to propose an inorganic abiogenic hypothesis for petroleum formation after he observed petroleum springs in the Bay of Cumaux (Cumana) on the northeast coast of Venezuela. In 1804 he is quoted as saying, "petroleum is the product of a distillation from great depth and issues from the primitive rocks beneath which the forces of all volcanic action lie." Abraham Gottlob Werner and the proponents of neptunism in the 18th century believed basaltic sills to be solidified oils or bitumen.

While these notions have been proven unfounded, the basic idea that petroleum is associated with magmatism persisted. Other prominent proponents of what would become the abiogenic hypothesis included Mendeleev and Berthelot.

Russian geologist Nikolai Alexandrovitch Kudryavtsev proposed the modern abiotic hypothesis of petroleum in 1951. On the basis of his analysis of the Athabasca Oil Sands in Alberta, Canada, he concluded that no "source rocks" could form the enormous volume of hydrocarbons, and that therefore the most plausible explanation is abiotic deep petroleum.

However, humic coals have since been proposed for the source rocks. Kudryavtsev's work was continued by Petr N. Kropotkin, Vladimir B. Porfir'ev, Emmanuil B. Chekaliuk, Vladilen A. Krayushkin, Georgi E. Boyko, Georgi I. Voitov, Grygori N. Dolenko, Iona V. Greenberg, Nikolai S. Beskrovny, and Victor F. Linetsky. Astronomer Thomas Gold was the most prominent proponent of the abiogenic hypothesis in the West until his death in 2004. Currently, Jack Kenney of Gas Resources Corporation is a prominent proponent in the West.

State of Current Research

Little research is directed on establishing abiogenic petroleum or methane, although the Carnegie Institution for Science have found that ethane and heavier hydrocarbons can be synthesized under conditions of the upper mantle. Research mostly related to astrobiology and the deep microbial biosphere and serpentinite reactions, however, continue to provide insight into the contribution of abiogenic hydrocarbons into petroleum accumulations.

- rock porosity and migration pathways for abiogenic petroleum
- ocean floor hydrothermal vents as in the Lost City hydrothermal field;
- Mud volcanoes and the volatile contents of deep pelagic oozes and deep formation brines

- mantle peridotite serpentinization reactions and other natural Fischer-Tropsch analogs
- Primordial hydrocarbons in meteorites, comets, asteroids and the solid bodies of the solar system
  - Primordial or ancient sources of hydrocarbons or carbon in Earth:
    - Primordial hydrocarbons formed from hydrolysis of metal carbides of the iron peak of cosmic elemental abundance (Cr, Fe, Ni, V, Mn, Co)
- isotopic studies of groundwater reservoirs, sedimentary cements, formation gases and the composition of the noble gases and nitrogen in many oil fields
- the geochemistry of petroleum and the presence of trace metals related to Earth's mantle (Ni, V, Cd, As, Pb, Zn, Hg and others).

Similarly, research into the deep microbial hypothesis of hydrocarbon generation is advancing as part of the attempt to investigate the concept of panspermia and astrobiology, specifically using deep microbial life as an analog for life on Mars. Research applicable to deep microbial petroleum theories includes

- Research into how to sample deep reservoirs and rocks without contamination
- Sampling deep rocks and measuring chemistry and biological activity
- Possible energy sources and metabolic pathways which may be used in a deep biosphere
- Investigations into the reworking primordial hydrocarbons by bacteria and their effects on carbon isotope fractionation.

A 2006 review article by Glasby presented arguments against the abiogenic origin of petroleum on a number of counts.

Foundations of the Hypotheses

Within the mantle, carbon may exist as hydrocarbons, chiefly methane and as elemental carbon, carbon dioxide and carbonates. The abiotic hypothesis is that the full suite of hydrocarbons found in petroleum can be generated in the mantle by abiogenic processes, and these hydrocarbons can migrate out of the mantle into the crust until they escape to the surface or are trapped by impermeable strata, forming petroleum reservoirs.

Abiogenic theories reject the supposition that certain molecules found within petroleum, known as biomarkers, are indicative of the biological origin of petroleum. They contend that these molecules mostly come from microbes feeding on petroleum in its upward migration through the crust, that some of them are found in meteorites, which have presumably never contacted living material, and that some can be generated abiogenically by plausible reactions in petroleum.

The hypothesis is founded primarily upon:

Proponents	***Item***
Gold	The presence of methane on other planets, meteors, moons and comets Gold,
Kenney	Proposed mechanisms of abiotically chemically synthesizing hydrocarbons within the mantle
Kudryavtsev, Gold	Hydrocarbon-rich areas tend to be hydrocarbon-rich at many different levels
Kudryavtsev, Gold	Petroleum and methane deposits are found in large patterns related to deep-seated large-scale structural features of the crust rather than to the patchwork of sedimentary deposits
Gold	Interpretations of the chemical and isotopic composition of natural petroleum
Kudryavtsev, Gold	The presence of oil and methane within non-sedimentary rocks upon the Earth
Gold	The existence of methane hydrate deposits
Gold	Perceived ambiguity in some assumptions and key evidence used in the orthodox biogenic petroleum theories
Gold	Bituminous coal creation is based upon deep hydrocarbon seeps
Gold	Surface carbon budget and oxygen levels stable over geologic time scales
Kudryavtsev, Gold	Biogenic theories do not explain some hydrocarbon deposit characteristics
Szatmari	The distribution of metals in crude oils fits better with upper serpentinized mantle, primitive mantle and chondrite patterns than oceanic and continental crust, and show no correlation with sea water
Gold	The association of hydrocarbons with helium, a noble gas

Conventional Theories

According to generally accepted theory, petroleum is derived from ancient biomass. The theory was initially based on the isolation of molecules from petroleum that closely resemble known biomolecules.

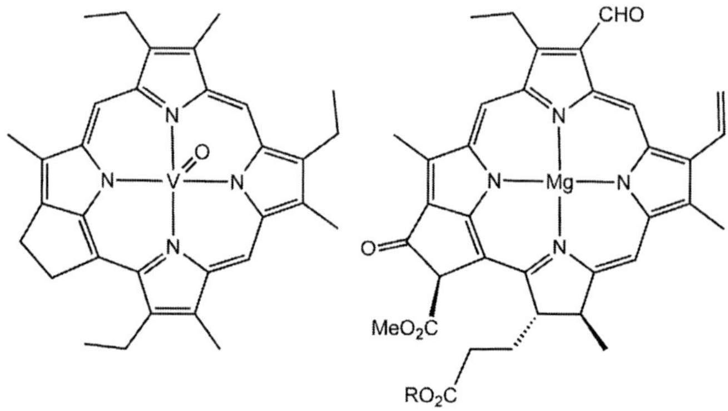

Figure 2: Structure of a biomarker extracted from petroleum and simplified structure of chlorophyll a.

Most petroleum geologists prefer theories of oil formation, which holds that oil originated in shallow seas as vast quantities of marine plankton or plant materials died and sank into the mud. Under the resulting anaerobic conditions organic compounds remained in a reduced state where anaerobic bacteria converted the lipids (fats, oils and waxes) into a waxy substance called kerogen.

As the source rock was buried deeper, overburden pressure raised temperatures into the *oil window*, between 80 and 180 °C. Most of the organic compounds degraded into the straight-chain hydrocarbons that comprise most of petroleum. This process is called the generation *kitchen*. Once crude oil formed, it became very fluid and migrated upward through the rock strata. This process is called oil expulsion. Eventually it was either trapped in an oil reservoir or oil escaped to the surface and was biodegraded by soil bacteria. Oil buried deeper entered the "gas window" of more than 160 °C and was converted into natural gas by thermal cracking. Thus, theory predicts that no oil will be found below a certain depth, only unassociated gas. At greater depths, even natural gas would be pyrolyzed.

Proposed Mechanisms of Abiogenic Petroleum

Primordial Deposits

Thomas Gold's work was focused on hydrocarbon deposits of primordial origin. Meteorites are believed to represent the major composition of material from which the Earth was formed. Some meteorites, such as carbonaceous chondrites, contain carbonaceous material. If a large amount of this material is still within the Earth, it could have been leaking upward for billions of years. The thermodynamic conditions within the mantle would allow many hydrocarbon molecules to be at equilibrium under high pressure and high temperature. Although molecules in these conditions may disassociate, resulting fragments would be reformed due to the pressure. An average equilibrium of various molecules would exist depending upon conditions and the carbon-hydrogen ratio of the material.

Creation within the Mantle

Russian researchers concluded that hydrocarbon mixes would be created within the mantle. Experiments under high temperatures and pressures produced many hydrocarbons, including n-alkanes through $C_{10}H_{22}$, from iron oxide, calcium carbonate, and water. Because such materials are in the mantle and in subducted crust, there is no requirement that all hydrocarbons be produced from primordial deposits.

Hydrogen Generation

Hydrogen gas and water have been found more than 6 kilometres deep in the upper crust, including in the Siljan Ring boreholes and the Kola Superdeep Borehole. Data from the western United States suggests that aquifers from near the surface may extend to depths of 10 to 20 km. Hydrogen gas can be created by water reacting with silicates, quartz and feldspar, in temperatures in the 25° to 270°C range. These minerals are common in crustal rocks such as granite. Hydrogen may react with dissolved carbon compounds in water to form methane and higher carbon compounds.

One reaction not involving silicates which can create hydrogen is:

$$\textit{Ferrous oxide} + \textit{Water} \rightarrow \textit{Magnetite} + \textit{hydrogen}$$

$$3FeO + H_2O \rightarrow Fe_3O_4 + H_2$$

The above reaction operates best at low pressures. At pressures greater than 5 GPa almost no hydrogen is created.

Serpentinite Mechanism

One proposed mechanism by which abiogenic petroleum is formed was first proposed by the Ukrainian scientist, Prof. Emmanuil B. Chekaliuk in 1967. He proposed that petroleum could be formed at high temperatures and pressures from inorganic carbon in the form of carbon dioxide, hydrogen and/or methane.

This mechanism is supported by several lines of evidence which are accepted by modern scientific literature. This involves synthesis of oil within the crust via catalysis by chemically reductive rocks. A proposed mechanism for the formation of inorganic hydrocarbons is via natural analogs of the Fischer-Tropsch process known as the *serpentinite mechanism* or the serpentinite process.

$$CH_4 + \frac{1}{2}O_2 \rightarrow 2H_2 + CO$$

$$(2n+1)H_2 + nCO \rightarrow C_nH_{2n+2} + nH_2O$$

Serpentinites are ideal rocks to host this process as they are formed from peridotites and dunites, rocks which contain greater than 80% olivine and usually a percentage of Fe-Ti spinel minerals. Most olivines also contain high nickel concentrations (up to several percent) and may also contain chromite or chromium as a contaminant in olivine, providing the needed transition metals.

However, serpentinite synthesis and spinel cracking reactions require hydrothermal alteration of pristine peridotite-dunite, which is a finite process intrinsically related to metamorphism, and further, requires significant addition of water. Serpentinite is unstable at mantle temperatures and is readily dehydrated to granulite, amphibolite, talc-schist and even eclogite.

This suggests that methanogenesis in the presence of serpentinites is restricted in space and time to mid-ocean ridges and upper levels of subduction zones. However, water has been found as deep as 12 km, so water-based reactions are dependent upon the local conditions. Oil being created by this process in intracratonic regions is limited by the materials and temperature.

Serpentinite Synthesis

A chemical basis for the abiotic petroleum process is the serpentinization of peridotite, beginning with methanogenesis via hydrolysis of olivine into serpentine in the presence of carbon dioxide. Olivine, composed of Forsterite and Fayalite metamorphoses into serpentine, magnetite and silica by the following reactions, with silica from fayalite decomposition (reaction 1a) feeding into the forsterite reaction (1b).

Reaction 1a:

Fayalite + water → Magnetite + aqueous silica + Hydrogen

$$3Fe_2SiO_4 + 2H_2O \rightarrow 2Fe_3O_4 + 3SiO_2 + 2H_2$$

Reaction 1b:

Forsterite + aqueous silica → Serpentinite

$$3Mg_2SiO_4 + SiO_2 + 4H_2O \rightarrow 2Mg_3Si_2O_5(OH)_4$$

When this reaction occurs in the presence of dissolved carbon dioxide (carbonic acid) at temperatures above 500 °C Reaction 2a takes place.

Reaction 2a:

Olivine + Water + Carbonic acid → Serpentine + Magnetite + Methane

$$(Fe,Mg)_2SiO_4 + nH_2O + CO_2 \rightarrow Mg_3Si_2O_5(OH)_4 + Fe_3O_4 + C$$

or, in balanced form:

$$18Mg_2SiO_4 + 6Fe_2SiO_4 + 26H_2O + CO_2 \rightarrow 12Mg_3Si_2O_5(OH)_4 + 4Fe_3O_4 + CH_4$$

However, reaction 2(b) is just as likely, and supported by the presence of abundant talc-carbonate schists and magnesite stringer veins in many serpentinised peridotites;

Reaction 2b:

Olivine + Water + Carbonic acid → Serpentine + Magnetite + Magnesite + Silica

$$(Fe,Mg)_2SiO_4 + nH_2O + CO_2 \rightarrow Mg_3Si_2O_5(OH)_4 + Fe_3O_4 + M$$

The upgrading of methane to higher n-alkane hydrocarbons is via dehydrogenation of methane in the presence of catalyst transition metals (e.g. Fe, Ni). This can be termed spinel hydrolysis.

Spinel Polymerization Mechanism

Magnetite, chromite and ilmenite are Fe-spinel group minerals found in many rocks but rarely as a major component in non-ultramafic rocks. In these rocks, high concentrations of magmatic magnetite, chromite and ilmenite provide a reduced matrix which may allow abiotic cracking of methane to higher hydrocarbons during hydrothermal events.

Chemically reduced rocks are required to drive this reaction and high temperatures are required to allow methane to be polymerized to ethane.

Note that reaction 1a, above, also creates magnetite.

Reaction 3:

Methane + Magnetite → Ethane + Hematite

$$nCH_4 + nFe_3O_4 + nH_2O \rightarrow C_2H_6 + Fe_2O_3 + HCO_3 + H^+$$

Reaction 3 results in n-alkane hydrocarbons, including linear saturated hydrocarbons, alcohols, aldehydes, ketones, aromatics, and cyclic compounds.

Carbonate Decomposition

Calcium carbonate may decompose at around 500 °C through the following reaction:

Reaction 5:

Hydrogen + Calcium carbonate → Methane + Calcium oxide + Water

$$4H_2 + CaCO_3 \rightarrow CH_4 + CaO + 2H_2O$$

Note that CaO (lime) is not a mineral species found within natural rocks. Whilst this reaction is possible, it is not plausible.

Evidence of Abiogenic Mechanisms

- Calculations by J.F. Kenney using scaled particle theory for a simplified perturbed hard-chain, (a statistical mechanical model) predict that methane compressed to 30 or 40 kbar at 1000 °C (conditions in the mantle) is relatively unstable in relation to higher hydrocarbons. However, these calculations do not include methane pyrolisis yielding amorphous carbon and hydrogen, which is recognized as the prevalent reaction at high temperatures.
- Experiments in diamond anvil high pressure cells have resulted in partial conversion of methane and inorganic carbonates into light hydrocarbons.

Biotic (Microbial) Hydrocarbons

The "deep biotic petroleum hypothesis", similar to the abiogenic petroleum origin hypothesis, holds that not all petroleum deposits within the Earth's rocks can be explained purely according to the orthodox view of petroleum geology. Thomas Gold used the term *the deep hot biosphere* to describe the microbes which live underground.

This hypothesis is different from biogenic oil in that the role of deep-dwelling microbes is a biological source for oil which is not of a sedimentary origin and is not sourced from surface carbon. Deep microbial life is only a contaminant of primordial hydrocarbons. Parts of microbes yield molecules as biomarkers.

Deep biotic oil is considered to be formed as a byproduct of the life cycle of deep microbes. Shallow biotic oil is considered to be formed as a byproduct of the life cycles of shallow microbes.

Microbial Biomarkers

Thomas Gold, in a 1999 book, cited the discovery of thermophile bacteria in the Earth's crust as new support for the postulate that these bacteria could explain the existence of certain biomarkers in extracted petroleum. Thorough rebuttal of biogenic origins based on biomarkers has been offered by Kenney, et al. (2001).

Isotopic Evidence

Methane is ubiquitous in crustal fluid and gas. Research continues to attempt to characterise crustal sources of methane as biogenic or abiogenic using carbon isotope fractionation of observed gases. There are few clear examples of abiogenic methane-ethane-butane, as the same processes favour enrichment of light isotopes in all chemical reactions, whether organic or inorganic. $\delta^{13}C$ of methane overlaps that of inorganic carbonate and graphite in the crust, which are heavily depleted in ^{12}C, and attain this by isotopic fractionation during metamorphic reactions.

One argument for abiogenic oil cites the high carbon depletion of methane as stemming from the observed carbon isotope depletion with depth in the crust. However, diamonds, which are definitively of mantle origin, are not as depleted as methane, which implies that methane carbon isotope fractionation is not controlled by mantle values. Commercially extractable concentrations of helium (greater than 0.3%) are present in natural gas from the Panhandle-Hugoton fields in the USA, as well as from some Algerian and Russian gas fields. Helium trapped within most petroleum occurrences, such as the occurrence in Texas, is of a distinctly crustal character with an

Ra ratio of less than 0.0001 that of the atmosphere. The Chimaera gas seep, near Antalya (SW Turkey), new and thorough molecular and isotopic analyses including methane (~87% v/v; D13C1 from -7.9 to -12.3 ‰; D13D1 from -119 to -124 ‰), light alkanes (C2+C3+C4+C5 = 0.5%; C6+: 0.07%; D13C2 from -24.2 to -26.5 ‰; D13C3 from -25.5 to -27 ‰), hydrogen (7.5 to 11 %), carbon dioxide (0.01-0.07%; D13CCO2: -15 ‰), helium (~80 ppmv; R/Ra: 0.41) and nitrogen (2-4.9%; D15N from -2 to -2.8 ‰) converge to indicate that the seep releases a mixture of organic thermogenic gas, related to mature Type III kerogen occurring in Paleozoic and Mesozoic organic rich sedimentary rocks, and abiogenic gas produced by low temperature serpentinization in the Tekirova ophiolitic unit.

Biomarker Chemicals

Certain chemicals found in naturally occurring petroleum contain chemical and structural similarities to compounds found within many living organisms. These include terpenoids, terpenes, pristane, phytane, cholestane, chlorins and porphyrins, which are large, chelating molecules in the same family as heme and chlorophyll. Materials which suggest certain biological processes include tetracyclic diterpane and oleanane.

The presence of these chemicals in crude oil is assumed to be as a result of the inclusion of biological material in the oil. This is predicated upon the theory that these chemicals are released by kerogen during the production of hydrocarbon oils, as these are chemicals highly resistant to degradation and plausible chemical paths have been studied.

Abiotic defenders state that biomarkers get into oil during its way up as it gets in touch with ancient fossils. However a more plausible explanation is that biomarkers are traces of biological molecules from bacteria (archaea) that feed on primordial hydrocarbons and die in that environment. For example, hopanoids are just parts of the bacterial cell wall present in oil as contaminant.

Trace Metals

Nickel (Ni), vanadium (V), lead (Pb), arsenic (As), cadmium (Cd), mercury (Hg) and others metals frequently occur in oils. Some heavy crude oils, such as Venezuelan heavy crude have up to 45% vanadium pentoxide content in their ash, high enough that it is a commercial source for vanadium. Abiotic supporters argue that these metals are common in Earth's mantle, but relatively high contents of nickel, vanadium, lead and arsenic can be usually found in almost all marine sediments.

Analysis of 22 trace elements in oils correlate significantly better with chondrite, serpentinized fertile mantle peridotite, and the primitive mantle than with oceanic or continental crust, and shows no correlation with seawater.

Reduced Carbon

Sir Robert Robinson studied the chemical makeup of natural petroleum oils in great detail, and concluded that they were mostly far too hydrogen-rich to be a likely product of the decay of plant debris, assuming a dual origin for Earth hydrocarbons. However, several processes which generate hydrogen could supply kerogen hydrogenation which is compatible with conventional petroleum generation theories.

Olefins, the unsaturated hydrocarbons, would have been expected to predominate by far in any material that was derived in that way. He also wrote: "Petroleum... [seems to be] a primordial hydrocarbon mixture into which bio-products have been added."

This has however been demonstrated later to be a misunderstanding by Robinson, related to the fact that only short duration experiments were available to him. Olefins are thermally very unstable (that is why natural petroleum normally does not contain such compounds) and in laboratory experiments that last more than a few hours, the olefins are no longer present.

The presence of low-oxygen and hydroxyl-poor hydrocarbons in natural living media is supported by the presence of natural waxes (n=30+), oils (n=20+) and lipids in both plant matter and animal matter, for instance fats in phytoplankton, zooplankton and so on. These oils and waxes, however, occur in quantities too small to significantly affect the overall hydrogen/carbon ratio of biological materials.

However, after the discovery of highly aliphatic biopolymers in algae, and that oil generating kerogen essentially represent concentrates of such materials, no theoretical problem exists anymore. Also, the millions of source rock samples that have been analyzed for petroleum yield by the petroleum industry have confirmed the large quantities of petroleum found in sedimentary basins.

Field Observations

Field observations sometimes cited as commercial occurrence of abiotic petroleum, but disputed by critics, include the Siljan Ring, offshore Vietnam, Eugene Island block 330 oil field, and the Dnieper-Donets Basin.

The Russian-Ukrainian school saw evidence of their theory in the fact that some oil reservoirs exist in non-sedimentary rocks such as granite, metamorphic or porous volcanic rocks. However, critics note that non-sedimentary rocks served as reservoirs for oil expelled from nearby sedimentary source rock through common migration or re-migration mechanisms.

The following observations have been commonly been used to argue for the abiogenic hypothesis, however all of these petroleum occurrences can also be fully explained by conventional petroleum formation theories.

Lost City Hydrothermal Vent Field

The Lost City Hydrothermal Vent Field was determined to have abiogenic hydrocarbon production. Proskurowski et al. wrote, "Radiocarbon evidence rules out seawater bicarbonate as the carbon source for FTT reactions, suggesting that a mantle-derived inorganic carbon source is leached from the host rocks. Our findings illustrate that the abiotic synthesis of hydrocarbons in nature may occur in the presence of ultramafic rocks, water, and moderate amounts of heat."

Siljan Ring, Sweden

The Siljan Ring meteorite crater, Sweden, was proposed by Thomas Gold as the most likely place to test the hypothesis because it was one of the few places in the world where the granite basement was cracked sufficiently (by meteorite impact) to allow oil to seep up from the mantle; furthermore it is infilled with a relatively thin veneer of sediment, which was sufficient to trap any abiogenic oil, but was modelled as not having been subjected to the heat and pressure conditions (known as the "oil window") normally required to create biogenic oil. However, some geochemists concluded by geochemical analysis that the oil in the seeps came from the organic-rich Ordivician Tretaspis shale, where it was heated by the meteorite impact.

The Gravberg-1 borehole penetrated 7,500 m, through the deepest rock in the Siljan Ring in which proponents had hoped to find hydrocarbon reservoirs. Some eight barrels of magnetite paste and hydrocarbon-bearing sludge were recovered from the well; Gold maintained that the hydrocarbons were chemically different from, and not derived from, those added to the borehole, but analyses showed that the hydrocarbons were derived from the diesel fuel-based drilling fluid used in the drilling. This well also sampled over 13,000 feet (4,000 m) of methane-bearing inclusions.

A second borehole, Stenberg-1, was drilled a few miles away, finding similar results, This time no diesel fuel-based drilling fluid was found.

Bacterial Mats

Direct observation of bacterial mats and fracture-fill carbonate and humin of bacterial origin in deep boreholes in Iran, Australia, Sweden and Canada are also taken as evidence for the abiogenic origin of petroleum

Example Proposed Abiogenic Methane Deposits

Panhandle-Hugoton field (Anadarko Basin) in the south-central United States is the most important gas field with commercial helium content. The B¡ch Hφ oil field in Vietnam has been proposed as an example of abiogenic oil because it is 4,000 m of fractured basement granite, at a depth of 5,000 m. However, others argue that it contains biogenic oil which leaked into the basement horst from conventional source rocks within the Cuu Long basin.

A major component of mantle-derived carbon is indicated in commercial gas reservoirs in the Pannonian and Vienna basins of Hungary and Austria.

Natural gas pools interpreted as being mantle-derived are the Shengli Field and Songliao Basin, northeastern China.

The Chimaera gas seep, near Çýralý, Antalya (southwest Turkey), has been continuously active for millennia and it is known to be the source of the first Olympic fire in the Hellenistic period.

On the basis of chemical composition and isotopic analysis, the Chimaera gas is said to be about half biogenic and half abiogenic gas, the largest emission of biogenic methane discovered; deep and pressurized gas accumulations necessary to sustain the gas flow for millennia, posited to be from an inorganic source, may be present. Local geology of Chimaera flames, at exact position of flames, reveals contact between serpentinized ophiolite and carbonate rocks. Fischer-Tropsch process can be suitable reaction to form hydrocarbons gases.

The Geological Argument for Abiogenic Oil

Given the known occurrence of methane and the probable catalysis of methane into higher atomic weight hydrocarbon molecules, the abiogenic hypothesis considers the following to be key observations in support;

- The serpentinite synthesis, graphite synthesis and spinel catalysation models prove the process is viable
- The likelihood that abiogenic oil seeping up from the mantle is trapped beneath sediments which effectively seal mantle-tapping faults
- Mass-balance calculations for supergiant oilfields which argue that the calculated source rock could not have supplied the reservoir with the known accumulation of oil, implying deep recharge (Kudryavtsev, 1951)
- The presence of hydrocarbons encapsulated in diamonds.

Incidental Evidence

The proponents of abiogenic oil use several arguments which draw on a variety of natural phenomena in order to support the hypothesis

- The modelling of some researchers which shows the Earth was accreted at relatively low temperature, thereby perhaps preserving primordial carbon deposits within the mantle, to drive abiogenic hydrocarbon production
- The presence of methane within the gases and fluids of mid-ocean ridge spreading centre hydrothermal fields.

The Geological Argument Against

Key arguments against chemical reactions, such as the serpentinite mechanism, as being the major source of hydrocarbon deposits within the crust are;

- The lack of available pore space within rocks as depth increases
  - This is contradicted by numerous studies which have documented the existence of hydrologic systems operating over a range of scales and at all depths in the continental crust.

- The lack of any hydrocarbon within the crystalline shield areas of the major cratons, especially around key deep seated structures which are predicted to host oil by the abiogenic hypothesis.
- Limited evidence that major serpentinite belts underlie continental sedimentary basins which host oil
- Lack of conclusive proof that carbon isotope fractionation observed in crustal methane sources is entirely of abiogenic origin (Lollar et al. 2006)
- Mass balance problems of supplying enough carbon dioxide to serpentinite within the metamorphic event before the peridotite is fully reacted to serpentinite
- Drilling of the Siljan Ring failed to find commercial quantities of oil, thus providing a counter example to Kudryavtsev's Rule and failing to locate the predicted abiogenic oil.
- Helium in the Siljan Gravberg-1 well was depleted in ^{3}He and not consistent with a mantle origin
  - The Gravberg-1 well only produced 84 barrels (13.4 m^3) of oil, which later was shown to derive from organic additives, lubricants and mud used in the drilling process.
- The distribution of sedimentary basins is caused by plate tectonics, with sedimentary basins forming on either side of a volcanic arc, which explains the distribution of oil within these sedimentary basins
- Kudryavtsev's Rule has been explained for oil and gas (not coal): Gas deposits which are below oil deposits can be created from that oil or its source rocks. Because natural gas is less dense than oil, as kerogen and hydrocarbons are generating gas the gas fills the top of the available space. Oil is forced down, and can reach the spill point where oil leaks around the edge(s) of the formation and flows upward. If the original formation becomes completely filled with gas then all the oil will have leaked above the original location.
- Ubiquitous presence of diamondoids in natural hydrocarbons such as oil, gas and condensates are composed of carbon from biological sources, unlike the carbon found in normal diamonds.

Arguments against the Incidental Evidence

- Gas ruptures during earthquakes are more likely to be sourced from biogenic methane generated in unconsolidated sediment from existing organic matter, released by earthquake liquefaction of the reservoir during tremors
- The presence of methane hydrate is arguably produced by bacterial action upon organic detritus falling from the littoral zone and trapped in the depth due to pressure and temperature
- The likelihood of vast concentrations of methane in the mantle is very slim, given mantle xenoliths have negligible methane in their fluid inclusions; conventional plate tectonics explains deep focus quakes better, and the extreme confining pressures invalidate the hypothesis of gas pockets causing quakes
- Further evidence is the presence of diamond within kimberlites and lamproites which sample the mantle depths proposed as being the source region of mantle methane (by Gold et al.).

Crude Oil

Crude Oil Reservoirs

Three conditions must be present for oil reservoirs to form: a source rock rich in hydrocarbon material buried deep enough for subterranean heat to cook it into oil; a porous and permeable reservoir rock for it to accumulate in; and a cap rock (seal) or other mechanism that prevents it from escaping to the surface.

Within these reservoirs, fluids will typically organize themselves like a three-layer cake with a layer of water below the oil layer and a layer of gas above it, although the different layers vary in size between reservoirs. Because most hydrocarbons are lighter than rock or water, they often migrate upward through adjacent rock layers until either reaching the surface or becoming trapped within porous rocks (known as reservoirs) by impermeable rocks above.

However, the process is influenced by underground water flows, causing oil to migrate hundreds of kilometres horizontally or even short distances downward before becoming trapped in a reservoir. When hydrocarbons are concentrated in a trap, an oil field forms, from which the liquid can be extracted by drilling and pumping.

The reactions that produce oil and natural gas are often modelled as first order breakdown reactions, where hydrocarbons are broken down to oil and natural gas by a set of parallel reactions, and oil eventually breaks down to natural gas by another set of reactions. The latter set is regularly used in petrochemical plants and oil refineries.

Wells are drilled into oil reservoirs to extract the crude oil. "Natural lift" production methods that rely on the natural reservoir pressure to force the oil to the surface are usually sufficient for a while after reservoirs are first tapped. In some reservoirs, such as in the Middle East, the natural pressure is sufficient over a long time. The natural pressure in many reservoirs, however, eventually dissipates. Then the oil must be pumped out using "artificial lift" created by mechanical pumps powered by gas or electricity. Over time, these "primary" methods become less effective and "secondary" production methods may be used.

A common secondary method is "waterflood" or injection of water into the reservoir to increase pressure and force the oil to the drilled shaft or "wellbore." Eventually "tertiary" or "enhanced" oil recovery methods may be used to increase the oil's flow characteristics by injecting steam, carbon dioxide and other gases or chemicals into the reservoir. In the United States, primary production methods account for less than 40% of the oil produced on a daily basis, secondary methods account for about half, and tertiary recovery the remaining 10%. Extracting oil (or "bitumen") from oil/tar sand and oil shale deposits requires mining the sand or shale and heating it in a vessel or retort, or using "in-situ" methods of injecting heated liquids into the deposit and then pumping out the oil-saturated liquid.

Unconventional Oil Reservoirs

Oil-eating bacteria biodegrade oil that has escaped to the surface. Oil sands are reservoirs of partially biodegraded oil still in the process of escaping and being biodegraded, but they contain so much migrating oil that, although most of it has escaped, vast amounts are still present—more than can be found in conventional oil reservoirs. The lighter fractions of the crude oil are destroyed first, resulting in reservoirs containing an extremely heavy form of crude oil, called crude bitumen in Canada, or extra-heavy crude oil in Venezuela. These two countries have the world's largest deposits of oil sands.

On the other hand, oil shales are source rocks that have not been exposed to heat or pressure long enough to convert their trapped hydrocarbons into crude oil. Technically speaking, oil shales are not always shales and do not contain oil, but are fined-grain sedimentary rocks containing an insoluble organic solid called kerogen. The kerogen in the rock can be converted into crude oil using heat and pressure to simulate natural processes. The method has been known for centuries and was patented in 1694 under British Crown Patent No. 330 covering, "A way to extract and make great quantities of pitch, tar, and oil out of a sort of stone." Although oil shales are found in many countries, the United States has the world's largest deposits.

Petroleum Industry

The petroleum industry is involved in the global processes of exploration, extraction, refining, transporting (often with oil tankers and pipelines), and marketing petroleum products. The largest volume products of the industry are fuel oil and petrol. Petroleum is also the raw material for many chemical products, including pharmaceuticals, solvents, fertilizers, pesticides, and plastics. The industry is usually divided into three major components: upstream, midstream and downstream. Midstream operations are usually included in the downstream category.

Petroleum is vital to many industries, and is of importance to the maintenance of industrialized civilization itself, and thus is critical concern to many nations. Oil accounts for a large percentage of the world's energy consumption, ranging from a low of 32% for Europe and Asia, up to a high of 53% for the Middle East. Other geographic regions' consumption patterns are as follows: South and Central America (44%), Africa (41%), and North America (40%). The world at large consumes 30 billion barrels (4.8 km^3) of oil per year, and the top oil consumers largely consist of developed nations. In fact, 24% of the oil consumed in 2004 went to the United States alone, though by 2007 this had dropped to 21% of world oil consumed.

In the US, in the states of Arizona, California, Hawaii, Nevada, Oregon and Washington, the Western States Petroleum Association (WSPA) represents companies responsible for producing, distributing, refining, transporting and marketing petroleum. This non-profit trade association was founded in 1907, and is the oldest petroleum trade association in the United States.

Bibliography

Aftalion, Fred: *A History of the International Chemical Industry*, University of Pennsylvania Press, Philadelphia, 1991.

Aitchison, Leslie: *A History of Metals*: Interscience Publishers, New York, 1960.

Barnes, Thomas G.: *Origin and Destiny of the Earth's Magnetic Field,* Creation Research Society, London, 1983.

Bates, Robert Latimer: *Geology of the Industrial Rocks and Minerals*, Dover Publications, New York, 1969.

Belknap, D.F.: *Sea-level Rise on the Maine Islands: Proceedings*, Sixth Island Institute Conference, Hurricane Island, Maine, Island Institute, Rockland, ME, 1988.

Berlin, G.L.: *Earthquakes and the Urban Environment,* CRC Press, Boca Raton, FL, 1980.

Binay Kumar and R.P. Tandon: *Advances in Technologically Important Crystals*, Macmillan Publishers India, Delhi, 2007.

Biyani, A.K.: *Dimensions of Himalayan Geology*, Satish Serial Pub, Delhi, 2007.

Brauer, G.: *Handbook of Preparative Inorganic Chemistry*, Academic Press, NY, 1963.

Bromley, Simon: *American Hegemony and World Oil: The Industry, the State System, and the World Economy*. University Park, Pa., 1991.

Brookins, Douglas G.: *Mineral and Energy Resources: Occurrence, Exploitation, and Environmental Impact*, Merrill Pub. Co., Columbus, 1990.

Bruce A. Kennedy: *Surface Mining*, Society for Mining, Metallurgy, and Exploration, Littleton, Colo, 1990.

Chivers, T.: *A Guide To Chalcogen-Nitrogen Chemistry*, World Scientific, Singapore, 2004.

Chowdhary, L.R.: *Petroleum Geology of the Cambay Basin, Gujarat, India,* Indian Petroleum, Delhi, 2004.

Cotton, C.A.: *Volcanoes as Landscape Forms*, Hafner Publishing. New York, 1999.

Donald D. Carr.: *Industrial Minerals and Rocks,* Society for Mining, Metallurgy, and Exploration, Littleton, Colo, 1994.

Doyle, F. L.: *Karst Hydrogeology,* The University of Alabama in Huntsville Press, Huntsville, Alabama, 1977.

Dunai, T.J.: *Cosmogenic Nucleides,* Cambridge University Press, U.K., 2010.

Eaton, Rev. S.J.M.: *Petroleum, a History of the Oil Region,* J.P. Skelly & Co., Philadelphia, 1866.

Edward W.: *United States Oil Policy and Diplomacy: A Twentieth-Century Overview.* West-port, Conn.: Greenwood Press, 1983.

Edwards, Denis: *Earth Revealing Earth Healing: Ecology and Christian Theology,* Collegeville: Liturgical Press, 2001.

Edwards, Edward: *Pattern and Design with Dynamic Symmetry,* NY: Dover 1967.

Folk, R.L.: *The Petrology of Sedimentary Rocks,* Hemphill Publishing Co., Austin, 2001.

Garverick, Linda: *Corrosion in the Petrochemical Industry.* Materials Park, Ohio: ASM International, 2000.

Glenn T. Trewartha: *Elements of Geography,* New York, McGraw Hill, 1941.

Gontz, A.M.: *Evolution of Seabed Pockmarks in Penobscot Bay, Maine,* University of Maine, Orono, 2011.

Greenwood, N. N.; & Earnshaw, A.: *Chemistry of the Elements,* Butterworth-Heinemann, Oxford, 1997.

Greenwood, Norman N.; Earnshaw, Alan: *Chemistry of the Elements,* Butterworth-Heinemann, Oxford, 1997.

Gurjar, Ram Kumar: *Geography of Water Resources,* Rawat, Delhi, 2008.

Hall, Anthony: *Igneous Petrology,* Longman, New York, 1996.

Hartman, Howard L.: *Introductory Mining Engineering,* Wiley, New York, c1987.

Heal, H. G.: *The Inorganic Heterocyclic Chemistry of Sulfur, Nitrogen, and Phosphorus,* Academic Press, London, 1980.

Holmgren, D. A., Moody, J. D.: *The Structural Setting for Giant oil and Gas Fields,* Tokyo, 1975.

John C. Grover.: *Volcanic Eruptions and Great Earthquakes,* Copyright Publishing Company, Australia, 1998.

Kesler, Stephen E.: *Mineral Resources, Economics, and the Environment,* Maxwell Macmillan International, New York, 1994.

LeRoy, L. W. and others: *Subsurface Eeology: Petroleum, Mining, Construction,* Colorado School of Mines, Golden, Colo., 1987.

Lovesun, V.J. ; P.K. Sen and Amalendu Sinha: *Exploration, Exploitation, Enrichment and Environment of Coastal Placer Minerals,* Macmillan Publishers India, Delhi, 2007.

Majumdar P.K. Basu: *Law of Mines and Minerals*, Universal Law, Delhi, 2011.

Malhotra, Raman: *Mines and Minerals Laws*, Global India, Delhi, 2011.

Manthorp, P.A.: *Estuarine Circulation and Sediment Transport in the Saco River estuary*, Boston University, Boston, MA, 1995.

Mary Lambert: *Crystal Energy*, Cico Books, Delhi, 2010.

Mason, B.: *Meteorites*, John Wiley & Sons, New York, 1962.

McLaurin, John J.: *Sketches in Crude-oil*, J. Horace McFarland Company, Harrisburg, 1896.

Michael Mugerwa: *Fuel Cell Systems*. New York: Plenum Press, 1993.

Miller, J.R.: *Process Geomorphology*, Wm.C. Brown Publishers, Dubuque, Iowa, 1995.

Mohamed, Abdel-Mohsen; El-Gamal, Maisa: *Sulfur Concrete for the Construction Industry: A Sustainable Development Approach*, J. Ross Publishing, Fort Lauderdale, 2010.

Nechamkin, Howard: *The Chemistry of the Elements*, McGraw-Hill, New York, 1968.

Nehb, Wolfgang; Vydra, Karel: *Ullmann's Encyclopedia of Industrial Chemistry*, Wiley-VCH Verlag, 2006.

Peters, William C.: *Exploration and Mining Geology*, J. Wiley, New York, 1987.

Poonam Johri: *Minerals and Water*, Sonali, Delhi, 2005.

Pramod O. Alexander: *A Handbook of Minerals, Crystals, Rocks and Ores*, New India Pub Agency, Delhi, 2009.

Prieto, Carlos: *Mining in the New World*, McGraw-Hill, New York, 1973.

Reiter, L.: *Earthquake Hazard Analysis: Issues and Insights*, Columbia University Press, New York, 1990.

Riegel, Emil; Kent, James: *Kent and Riegel's Handbook of Industrial Chemistry and Biotechnology*, Springer, New York, 2007.

Roberts, Willard Lincoln, Thomas J. Campbell, and George Robert Rapp.: *Encyclopedia of Minerals*, Van Nostrand Reinhold, New York, 1990.

Samuel Delvin: *Vitamins, Minerals and Hormones*, Ivy Pub, Delhi, 2003.

Sawant, P. T.: *Engineering and General Geology*, New India Publishing Agency, Delhi, 2011.

Scalisi, Philip, and David Cook: *Classic Mineral Localities of the World, Asia and Australia*, Van Nostrand Reinhold Co., New York, 1983.

Selby, Michael John: *Earth's Changing Surface: an Introduction to Geomorphology*, Clarendon Press, Oxford, 1985.

Shepherd, R.: *Ancient Mining*, Elsevier Applied Science, London, New York, c1993.

Sidney Russell Cooke: *The Oil Trusts and Anglo-American Relations.* Macmillan, 1924

Steven, A.: *Grand Canyon: Monument to Catastrophe,* Institute for Creation Research, London,1994.

Summerfield, M.A.: *Global Geomorphology*, Pearson Education Ltd, Delhi, 1991.

Tiwari, S.K.: *Ore Geology, Economic Minerals And Mineral Economics*, Atlantic Publishers, Delhi, 2010.

Tracy, Robert J.; Owens, Brent: *Petrology: Igneous, Sedimentary, and Metamorphic,* New York: W. H. Freeman, 2005.

Vaughan, D. J.; Craig, J. R.: *Mineral Chemistry of Metal Sulfides*, Cambridge University Press, Cambridge, 1978.

Vijayalakshmi, G S: *Fuels and Biofuels*, Agrobios, Delhi, 2007.

Wahab, M. A.: *Essentials of Crystallography*, Narosa, Delhi, 2009.

White, William B.: *Geomorphology and Hydrology of Karst Terrains,* Oxford University Press, New York, 1988.

Whitmore, G: *Chemical Crystallography And Lipid Crystals*, Sarup Book, Delhi, 2009.

Wolfe, John A: *Mineral Resources: A World Review*, Chapman & Hall, New York, 1984.

Young, George J.: *Fuel Cells,* New York: Reinhold Publishing Corp., 1961.

Index

❑❑❑